Learning
MATHEMATICS
— The Fun Way

Mathematical Gym for Young Mathematicians

Authored by
Sumita Bose

V&S PUBLISHERS

Published by:

V&S PUBLISHERS

F-2/16, Ansari road, Daryaganj, New Delhi-110002
☎ 23240026, 23240027 • *Fax:* 011-23240028
Email: info@vspublishers.com • *Website:* www.vspublishers.com

Regional Office : Hyderabad
5-1-707/1, Brij Bhawan (Beside Central Bank of India Lane)
Bank Street, Koti, Hyderabad - 500 095
☎ 040-24737290
E-mail: vspublishershyd@gmail.com

Branch Office : Mumbai
Jaywant Industrial Estate, 1st Floor–108, Tardeo Road
Opposite Sobo Central Mall, Mumbai – 400 034
☎ 022-23510736
E-mail: vspublishersmum@gmail.com

BUY OUR BOOKS FROM: AMAZON FLIPKART

© **Copyright: Author**
ISBN 978-93-815888-0-2
Edition 2020

PUBLISHER'S NOTE

After publishing a series of interesting books on Mathematics for children, such as *Quiz Time Mathematics*, *Mathemagic Puzzles & Brain Drainers*, **V&S Publishers** has now come up with an altogether different and fun-filled book full of games, activities, puzzles, practice test papers and worksheets based on simple principles of Mathematics for kids till the age of eight or nine.

The book teaches the young learners the various branches of Mathematics like Number System, Addition, Subtraction, Multiplication and Division with numbers, Shapes and Sizes of objects present in the Nature, making them aware of the concepts of Length, Breadth, Height, Weight, Capacity, etc. and their measurements in a simple and interesting manner.

This book will be of great help to the teachers, especially of the primary sections of schools and act as a useful handbook for parents at home, who are always very keen on finding ways and means to teach and guide their young ones on the various subjects studied at school in a simple, fun-filled and effective manner.

The book actually makes Mathematics really easy and fun to learn, solve and understand.

ACKNOWLEDGEMENTS

"Where the clear stream of reason has not lost its way

Into the dreary desert sand of dead habit

Where the mind is led forward by thee

Into ever-widening thought and action

Into that heaven of freedom, my Father, let my country awake"

— **Rabindranath Tagore**

My heartfelt gratitude to my mother and family members for lighting my way with soothing light of encouragement.

To all my students and contemporary teachers my heartiest wishes for keeping alight my passion for knowledge.

Finally, I am thankful to Mr. Sahil Gupta and his super team who toiled night and day to make this book a success.

CONTENTS

PREFACE

Trends and buzzwords come and go in education but the need for differential instruction is constant — the students deserve to have their individual learning needs satisfied.

Learning Mathematics the Fun Way adheres to the vision of the latest National Curriculum Framework. It relates mathematical concepts to the child's own environment and experiences thereby inculcating the skills of quantifying a physical experience.

The main aim is to hold a modern approach but flexibly designed to keep the fun quotient high, so that learning becomes synonymous to enjoyment.

Each chapter starts by focusing on clear objectives — interesting but easy activities, games, puzzles and self-tests — to finally come upon the acquired skill by the end of the chapter. The common mistakes committed by the students are pointed out in each chapter along with the correct solutions. The approach is based on fostering cognitive skills leading to creativity betterment and increases the motivation to learn by keeping things simple.

The lucid language helps the children understand every chapter on their own. It is suggested that only when the student assimilates the content of one chapter the next chapter is attempted. The advantage of this book is that it emphasises on teaching with meaning, drawing on real-life models to help students gain confidence, enhance problem solving skills and imbibe a sense of achievement.

Taking advantage of my long experience in teaching Mathematics my aim has been to showcase that acquiring skills in numbers can be interesting, relevant and importantly super fun!

This book is a creation not only of the author but also of the editor, Mr. D J Borah as well as the designing and production team at V&S Publishers. The guidance from Mr. Sahil Gupta is incomparable. I sincerely thank them all for their valuable inputs.

As there is always a scope for improvement, any suggestion from my fellow teachers, parents and students are most welcome.

— **Sumita Bose**

NUMBER SYSTEM

Learning Objective:

To compare three-digit numbers and to find the place value of a digit in a three-digit number.

Number of Players:

Two

Materials Required:

White paper, thin cardboard, glue, a pair of scissors, sketch pens

Procedure:

Number Cards

Smallest 3-digit number	Smallest 3-digit number using digits 8, 4 and 6
Largest 3-digit number using digits 3, 0 and 1	Which is greater: 589 or 598?
Place Value of 6 in 369	Largest 3-digit number

Number Game Cards

- ☞ Write numbers 1 to 10 on a white sheet of paper using sketch pens.
- ☞ Cut and paste them on a thin cardboard.
- ☞ Make three sets of number cards for each player.
- ☞ Cut out the number game cards from the back of the book.
- ☞ Shuffle the number game cards and put them face down (written side down).
- ☞ Each player picks up a number game card by turn and reads the question on the card.
- ☞ The players write the answer by arranging their number cards.
- ☞ The player with the first correct answer scores two points.
- ☞ At the end of six rounds, the total score is calculated.
- ☞ The player with the higher total wins the game.

Example:

- ✶ Each player takes three sets of number cards.
- ✶ The players pick up a number game card by turn.
- ✶ Suppose the first player picks up the following card:

> Which is greater:
> 362 or 632?

- ✶ The player who arranges the number cards in the following manner first, scores two points.

- ✶ The game continues for six rounds.
- ✶ The player with the higher total score wins the game.

Skill Developed:

Comparison of 3-digit numbers, place value of digits in a 3-digit number.

Learning Mathematics - The Fun Way

Learning Objective:

To understand the place value of different digits in a three-digit number.

Materials Required:

Hundreds block, tens block, ones block, a pair of scissors, glue, white sheet of paper, pencil.

Procedure:

☞ Write HTO on top of the three-digit numbers: 145, 357, 289.

☞ Cut the number of ones block, tens block and hundreds block as per HTO of the number.

☞ Paste them on the paper and fill in the observation table.

Example:

✶ Write HTO on top of the number 236.

✶ H T O
   2 3 6

✶ Cut two hundreds block, three tens block and six ones block.

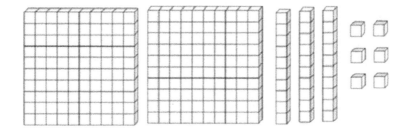

236 = 2 Hundreds, 3 Tens, 6 Ones

Observation:

145 = _ Hundreds, _ Tens, _ Ones

357 = _ Hundreds, _ Tens, _ Ones

289 = _ Hundreds, _ Tens, _ Ones

Result:

Every three-digit number has a ones digit, tens digit and a hundreds digit.

Puzzle

I am a three-digit number less than three centuries, but greater than two centuries. My one's digit is the sum of the ten's and hundred's digit. My ten's digit is an odd number less than five. Which number am I?

Common Error & Correction:

Error

 Place value of 3 in 387 = Hundreds

Correction

Place value of 3 in 387 = 3 Hundreds or 300

Tricks & Shortcuts:

There is often a confusion between the less than and greater than sign. An easy method to remember is:

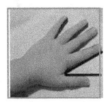

Left hand denotes the less than (<) sign.

WORKSHEET

Q1. Write the numbers and their names represented by the following blocks and notes:

a)

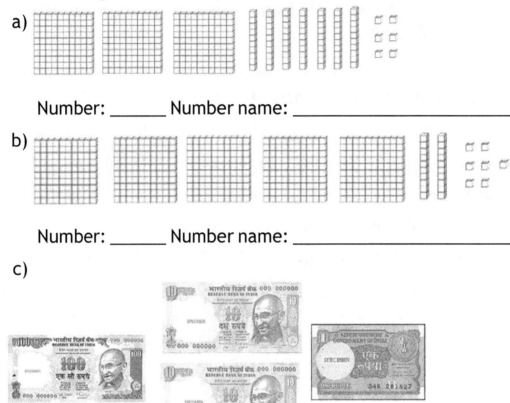

Number: _____ Number name: _____

b)

Number: _____ Number name: _____

c)

Number: _____ Number name: _____

Q2. Write the place values:

a)

    H   T   O

The place value of two is _____.

b)

    H   T   O

The place value of four is _____.

Q3. Colour the numbers as per the instructions given below:

A	369	804	212	114
B	100 + 10	873	110	378
C	800 + 9	890	810	801

a) Colour the smallest number in row A.

b) Colour two equal numbers in row B.

c) Colour the largest number in row C.

Q4. Complete the number patterns on the centipedes.

a)

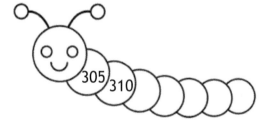

b)

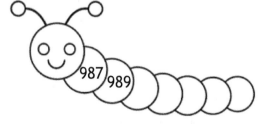

c)

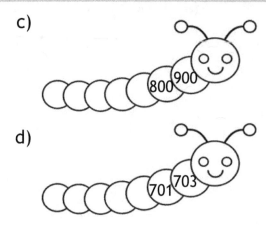

800 900

d)

701 703

Q5. Primary school students of different schools had to use buttons for doing an activity (Each student needs one button). Find out how many sheet of hundred buttons, sheet of ten buttons and loose items are required by different schools.

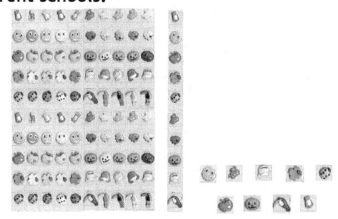

Sheet of hundred buttons   *Sheet of ten buttons*   *Loose items*

School	Number of Students	Sheet of hundred buttons	Sheet of ten buttons	Loose items
A	627			
B	418			
C	256			
D	563			

Learning Mathematics - The Fun Way

Q6. Solve the following Sudoku. Every row (left to right), every column (top to bottom) and every small square must have the numbers 1 to 4.

a)

1	3	4	
	2		3
2			
	4		

b)

	2	4	
1			3
4			2
	1	3	

Time: 45 minutes   Maximum marks: 10

Q1. Write the number names: (2 Marks)

   a. 948 _____

   b. 296 _____

Q2. Write the numeral: (2 Marks)

   a. Four hundred thirty five _____

   b. 700 + 60 + 2 = _____

Q3. Write the place value of one in the following numbers:

                                                                   (2 Marks)

	Number	Place Value
a.	133	_____
b.	216	_____

Q4. Fill in the blanks: (4 Marks)

   a. 891 _____ 819 (Put < or >)

   b. The smallest three-digit number using the digits 6, 0, 2 is
      _____.

   c. 340, 360, _____, _____. (Write the missing numbers)

   d. 563 = _ Hundreds + _ Tens + _ Ones

Learning Mathematics - The Fun Way

ADDITION

Learning Objective:

To develop strategic thinking along with addition.

Number of Players:

Two

Materials Required:

White paper, thin cardboard, glue, a pair of scissors, sketch pens

Procedure:

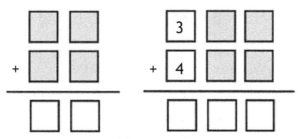

Addition Cards

☞ Make number cards similar to the game in the number chapter.

☞ Cut out the addition cards from the back of the book.

☞ Shuffle the two sets of number cards and put them face down.

☞ Give three sets of addition cards to each player.

☞ Players take turns turning over a number card and saying that number.

☞ The players can put the number cards anywhere in the blue boxes (one addition card is to be taken at a time).

☞ When all the four boxes are filled, the players find the sum.

☞ The player with the higher sum gets five points.

☞ At the end of five or six rounds, the points are added.

☞ The player with a higher total wins the game.

Example:

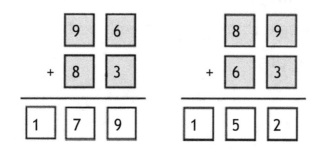

Player 1   Player 2

✭ Players take turns in turning the number cards.

✭ Suppose the turned cards are 9, 8, 6 and 3.

✭ The players arrange the cards as above (They can put the numbers anywhere in the blue boxes).

✭ The first player gets higher sum, hence gets five points.

✭ The game continues for five rounds.

✭ The player with a higher total wins the game.

Skill Developed:

Strategic thinking and addition of two and three-digit numbers.

Learning Objective:

To add three-digit numbers using hundreds, tens and ones blocks.

Materials Required:

Hundreds block, tens block, ones block, a pair of scissors, drawing sheet.

Procedure:

☞ Cut the number of ones block, tens block and hundreds block as per the given three-digit numbers.

☞ Arrange them on a drawing sheet.

☞ Next put all the ones block, all the tens block and all the hundreds block together.

☞ If each set is less than ten then the total number of blocks in each set represents the sum.

☞ If ones block are more than ten then regroup it to the tens block. Similarly, if the tens block are more than ten, then regroup it to the hundreds block.

☞ Count the number of hundreds blocks, tens blocks and ones blocks after regrouping. This gives the sum.

☞ Using the above method, find the sum of 253 and 311, 128 and 224.

☞ Fill in the observation table.

Example:

Q. Add: 341 + 286

✳ For 341, take 3 hundreds blocks, 4 tens blocks and 1 ones block.

341

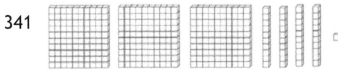

Addition

* For 286 take 2 hundreds blocks, 8 tens blocks and 6 ones blocks.

286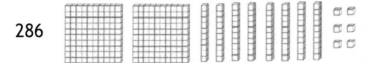

* To add 341 and 286, put all the ones blocks, all the tens blocks and all the hundreds blocks together.
* We get 5 hundreds blocks, 12 tens blocks and 7 ones blocks.

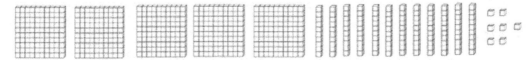

* Since tens block is more than ten, regroup it to hundreds. Hence we will get 6 hundreds, 2 tens and 7 ones. (Out of 12 tens, 10 tens combine to form 1 hundred).

Observation:

Problem	Addend	Addend	Sum
341 + 286	341	286	627

Result:

The three-digit numbers can be added by using hundreds, tens and ones blocks.

Learning Mathematics - The Fun Way

Deepayan made a magic card for his mother's thirtieth birthday. In the card, all the rows (left to right), all the columns (top to bottom) and both the diagonals (corner to corner) add up to thirty. Can you make a card which adds up to thirty-five?

8	11	10	1
9	2	7	12
3	12	9	6
10	5	4	11

(Hint: Try changing the numbers in the yellow box)

Common Error & Correction:

Error

$$
\begin{array}{r}
1\ 1 \\
2\ 7\ 8 \\
+\ 3\ 6\ 9 \\
\hline
5\ 3\ 7 \\
\hline
\end{array}
$$

Correction

$$
\begin{array}{r}
1\ 1 \\
2\ 7\ 8 \\
+\ 3\ 6\ 9 \\
\hline
6\ 4\ 7 \\
\hline
\end{array}
$$

Tricks & Shortcuts:

Add the digits of both the addends (the numbers being added) separately till you reach a single digit number. Do the same for the sum (answer).

Now add the single digits of the addends. If this sum is equal to the single digit of the sum, then the addition is correct.

Example:

$$
\begin{array}{r}
869 \\
+\ \ 107 \\
\hline
976
\end{array}
$$

869 Addend (8 + 6 + 9 = 23, 2 + 3 = 5)

+ 107 Addend (1 + 0 + 7 = 8)

976 Sum   (9 + 7 + 6 = 22, 2 + 2 = 4)

5 + 8 = 13, 1 + 3 = 4 same as the single digit obtained from the sum. Hence, the addition is correct.

WORKSHEET

Q1. Fill in the blanks:

a) 88 + 0 = _____

b) 37 + 98 = 98 + _____

c) 20 and 40 more is _____.

d) The sum of 8 and 42 is _____.

e) 15 more than 60 is _____.

Q2. Add the following numbers by using the number grid (Mark the arrows on the grid to show the addition).

a) 79 + 8 = _____

b) 28 + 60 = _____

c) 34 + 51 = _____

1	2	3	4	5	6	7	8	9	10
11	12	13	14	15	16	17	18	19	20
21	22	23	24	25	26	27	28	29	30
31	32	33	34	35	36	37	38	39	40
41	42	43	44	45	46	47	48	49	50
51	52	53	54	55	56	57	58	59	60
61	62	63	64	65	66	67	68	69	70
71	72	73	74	75	76	77	78	79	80
81	82	83	84	85	86	87	88	89	90
91	92	93	94	95	96	97	98	99	100

Learning Mathematics - The Fun Way

Q3. Write the numbers at the centre as the sum of two numbers (Write in five different ways).

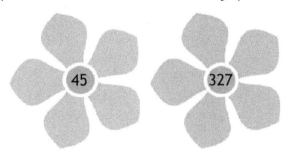

45

327

Q4. Add the following numbers by breaking them (i.e., by using expanded notation):

a) 83 + 12

= ☐ + ☐ + ☐ + ☐

= ☐ + ☐ + ☐ + ☐

= ☐ + ☐

= ☐

b) 243 + 131

= ☐ + ☐ + ☐ + ☐ + ☐ + ☐

= ☐ + ☐ + ☐ + ☐ + ☐ + ☐

= ☐ + ☐ + ☐

= ☐

Q5. Estimate the sum by rounding each addend to the nearest hundred.

a)   284 ⟶ _____
   + 576 ⟶ + _____

b)   321 ⟶ _____
   + 692 ⟶ + _____

c)   869 ⟶ _____
   + 142 ⟶ + _____

Addition

Q6. Find the sum:

a) 50 + 20 = _____

c) 123 + 321 = _____

b) 320 + 245 = _____

d) 9 + 41 + 100 = _____

Q7. Add:

a) 489
  + 268
  _____

b) 356
  + 507
  _____

c) 609
  + 195
  _____

d) 568
  + 158
  _____

Q8. In a Diwali fair, there were 408 children and 293 adults. How many people were there in all?

Q9. In an orchard, there are 359 orange trees and 591 mango trees. What is the total number of trees in the orchard?

Q10. Make your own three-digit addition problem and solve it. Use the following concept —boys and girls in a school.

Learning Mathematics - The Fun Way

PRACTICE TEST PAPER

Time: 45 minutes   Maximum marks: 10

Q1. Add the following numbers by breaking them (i.e. by using expanded notation): **(2 Marks)**

62 + 17

= ☐ + ☐ + ☐ + ☐

= ☐ + ☐ + ☐ + ☐

= ☐ + ☐

= ☐

Q2. Add the following numbers by using the number grid (Mark the arrows on the grid to show the addition). **(2 Marks)**

a. 13 + 50 = _____   b. 34 + 16 = _____

1	2	3	4	5	6	7	8	9	10
11	12	13	14	15	16	17	18	19	20
21	22	23	24	25	26	27	28	29	30
31	32	33	34	35	36	37	38	39	40
41	42	43	44	45	46	47	48	49	50
51	52	53	54	55	56	57	58	59	60
61	62	63	64	65	66	67	68	69	70
71	72	73	74	75	76	77	78	79	80
81	82	83	84	85	86	87	88	89	90
91	92	93	94	95	96	97	98	99	100

Addition

Q3. Fill in the blanks: (2 Marks)

   a. 11 more than 389 is _____.

   b. 404 + _____ = 404

Q4. In a library, there are 549 English books and 234 Hindi books. What is the total number of books in the library?

Q5. Complete the following addition crossword: (2 Marks)

49	+	76	=	
+		+		+
223	+		=	582
=		=		=
	+	435	=	

SUBTRACTION

Mathematical Magic

Learning Objective:

To show a mathematical magic based on subtraction by giving proper sequence of instructions.

Number of Players:

Two

Materials Required:

Paper and pencil

Procedure:

☞ Ask your friend to write any three-digit number in which the digits are consecutive (one after the other) and in the decreasing order from hundred's place to one's place. For example, 321.

☞ Ask him/her to reverse the number.

☞ Next ask him/her to subtract this new number from the original number.

☞ You can look at his/her forehead and at once tell the answer.

Example:

★ Suppose your friend chooses 876.

★ His/her reversed number would be 678.

★ He/she would work as follows:

$$
\begin{array}{r}
876 \\
-\ 678 \\
\hline
198
\end{array}
$$

Trick:

The answer is always 198 (Don't show the magic to the same person twice).

Skill Developed:

Subtraction of three-digit numbers.

Learning Objective:

To subtract three-digit numbers using hundreds, tens and ones block.

Materials Required:

Hundreds blocks, tens blocks, ones blocks, a pair of scissors, drawing sheet.

Procedure:

☞ Cut the number of ones blocks, tens blocks and hundreds blocks as per the given three-digit numbers.

☞ Arrange them on a drawing sheet.

☞ Next take away the ones blocks, tens blocks and hundreds blocks according to the subtrahend (the number being subtracted).

☞ The remaining number of hundreds blocks, tens blocks and ones blocks gives the difference.

☞ If the ones cannot be taken away, then regroup the tens and if the tens cannot be taken away, then regroup the hundreds and then take away the blocks.

☞ The remaining blocks gives the difference.

☞ Using the above method, find the difference of: a) 392 − 161 b) 427 − 258 c) 518 − 347

☞ Fill in the observation table.

Example:

Q. Subtract: 345 − 156.

★ For 345, take 3 hundreds blocks, 4 tens blocks and 5 ones blocks.

345 =

✳ For 156, take 1 hundreds block, 5 tens blocks and 6 ones blocks.

156 =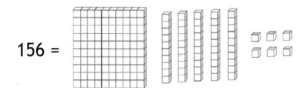

✳ We cannot take away 6 ones blocks from 5 ones blocks. Hence, we have to break one tens block of 345 into ones. So we get, 3 hundreds blocks, 3 tens blocks and 15 ones blocks.

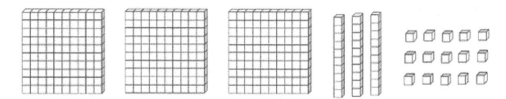

✳ Now we can take away the ones block (6 from 15), but still we cannot take away the tens. To do that, break one hundreds into tens. So we get, 2 hundreds blocks, 13 tens blocks and 15 ones blocks.

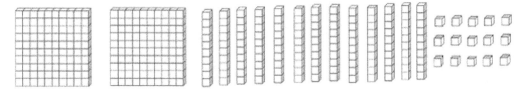

✳ Now we can subtract 156, i.e., take away 1 hundreds block, 5 tens blocks and 6 ones blocks.

✳ So we are left with 1 hundreds block, 8 tens blocks and 9 ones blocks.

189 =

Observation:

Problem	Minuend	Subtrahend	Difference
345 − 156	345	156	189

Result:

Three-digit numbers can be subtracted by using hundreds, tens and ones blocks.

Puzzle

A hunter saw four birds on a tree. He shot one of them. How many remained?

_____

Common Error & Correction:

Error	Correction

Error

```
      405
😞  − 2 8 3
    ─────
      2 8 2
```

Correction

```
      3 1
      ̸4 0 5
😊  − 2 8 3
    ─────
      1 2 2
```

Tricks & Shortcuts:

Add the difference to the subtrahend (the number being subtracted). If the answer is the minuend (the number from which it is subtracted), the subtraction is correct.

Example:

```
  786 Minuend
− 498 Subtrahend
─────────────
  288 Difference
```

Checking:
```
  288 Difference
+ 498 Subtrahend
─────────────
  786 Minuend
```

WORKSHEET

Q1. Subtract the following by counting up:

a) 58 − 23 = _____

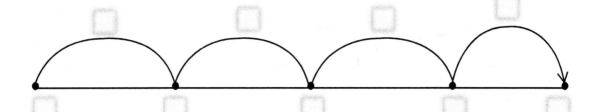

b) 80 − 55 = _____

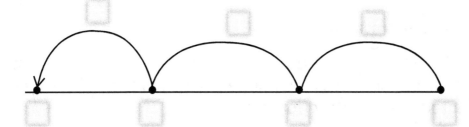

Q2. Fill in the blanks:

a) 989 − 0 = _____

b) 436 − 436 = _____

c) Reducing 87 by 37 gives _____.

d) Taking 29 away from 40, we get _____.

e) 300 less than 700 is _____.

Learning Mathematics - The Fun Way

Q3. Subtract the following numbers by using the number grid (Mark the arrows on the grid to show the subtraction).

1	2	3	4	5	6	7	8	9	10
11	12	13	14	15	16	17	18	19	20
21	22	23	24	25	26	27	28	29	30
31	32	33	34	35	36	37	38	39	40
41	42	43	44	45	46	47	48	49	50
51	52	53	54	55	56	57	58	59	60
61	62	63	64	65	66	67	68	69	70
71	72	73	74	75	76	77	78	79	80
81	82	83	84	85	86	87	88	89	90
91	92	93	94	95	96	97	98	99	100

a) $73 - 20 =$ _____   b) $88 - 63 =$ _____

Q4. Subtract the numbers connected by arrows in the subtraction cone. (Subtract the smaller number from the bigger number).

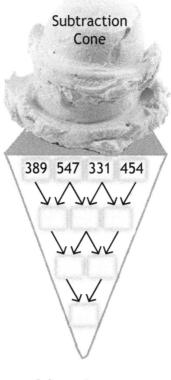

Subtraction Cone

389   547   331   454

Subtraction

37

Q5. Savitri Devi made 705 *rakhis*. She sold 596 of them. How many are not sold?

Q6. In Pakshi bird sanctuary, there are 834 birds, whereas in the Pakhi bird sanctuary, there are 843 birds. Which sanctuary has more birds? How many more?

PRACTICE TEST PAPER

Time: 45 minutes   Maximum marks: 10

Q1. Subtract the following numbers by using the number grid (Mark the arrows on the grid to show the subtraction).

(2 Marks)

1	2	3	4	5	6	7	8	9	10
11	12	13	14	15	16	17	18	19	20
21	22	23	24	25	26	27	28	29	30
31	32	33	34	35	36	37	38	39	40
41	42	43	44	45	46	47	48	49	50
51	52	53	54	55	56	57	58	59	60
61	62	63	64	65	66	67	68	69	70
71	72	73	74	75	76	77	78	79	80
81	82	83	84	85	86	87	88	89	90
91	92	93	94	95	96	97	98	99	100

a. 88 − 40 = _____   b. 63 − 22 = _____

Q2. Subtract: (2 Marks)

a. 765   b. 800
− 431   − 679
_____   _____

Q3. Fill in the blanks: (2 Marks)

a. 27 less than 438 is _____.

b. 898 − _____ = 898

Subtraction   39

Q4. In a box, there were 451 mangoes, out of which, 289 were distributed among children. How many mangoes were left in the box? **(2 Marks)**

Q5. A well is nine feet deep. A frog jumps up seven feet from the bottom of the well during day and falls down four feet at night. In how many days will he be able to jump out of the well? **(2 Marks)**

Learning Mathematics - The Fun Way

MULTIPLICATION

Learning Objective:

To develop the skill of learning multiplication facts.

Number of Players:

Two to four

Materials Required:

A deck of playing cards

Procedure:

☞ Remove all the jacks, queens and kings from the playing card set (The joker is considered as zero and ace as one).

☞ Keep the remaining cards face down at the centre.

☞ Each player picks two cards and speaks out the product by turn.

☞ The player with the highest product wins all the cards from the other players.

☞ If two players get the same product or any player gives the wrong answer, then those cards are returned to the centre.

☞ The game continues for seven or eight rounds.

☞ The player with the maximum number of cards at the end of eight rounds wins the game.

Example:

✶ Suppose the two cards turned by the first player are six and ace (Colour does not matter). Then the player will have to find the product 6 × 1.

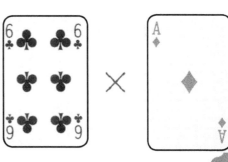

- ✻ The second player gets seven and eight, so he/she has to find 7× 8.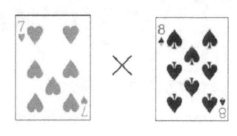
- ✻ Since fifty-six (7× 8) is greater than six (6 × 1), the second player will get all the four cards.
- ✻ The game will continue for eight rounds.
- ✻ The player with the maximum number of cards at the end of eight rounds will be the winner.

Skill Developed:

Multiplication facts (tables) from 2 to 10.

Activity

Learning Objective:

To develop the skill of learning multiplication facts by counting squares.

Materials Required:

Square paper, pencil, crayons.

Procedure:

☞ Take a ten by ten square paper and write 1 to 10 on top as well as left side (you may also take ten, ten by ten square papers instead of one and colour one fact on each sheet).

☞ For 2 × 1, colour two columns and one row, for 2 × 2, colour two columns and two rows, for 2 × 3, colour two columns and three rows.

☞ Count the number of boxes coloured in each figure. Continue the process till two columns and ten rows are coloured and complete the observation table.

Example:

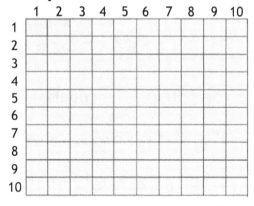

Observation: 2 × 1 = 2

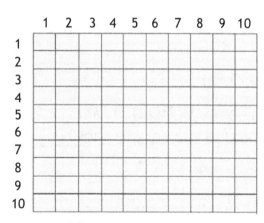

Observation: 2 × 2 = 4

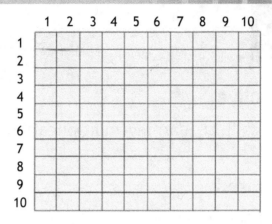

Observation: 2 × 3 = 6

Observation Table:

2 × 1 = 2,

2 × 2 = 4,

2 × 3 = -----

2 × 4 = -----

2 × 5 = ----

2 × 6 = ----

2 × 7 = ----

2 × 8 = ----

2 × 9 = ----

2 × 10 = ----

Result:

The multiplication facts of 2 to 10 can be obtained by counting the boxes on a square sheet.

Extension:

Write the multiplication facts of 3, 4, 5, 6, 7, 8, 9 and 10 by counting the boxes on a square paper.

Learning Mathematics - The Fun Way

Puzzle

When there are two flowers, the numbers are same. When there is a flower and a leaf, the numbers are different. Find the numbers.

$$\text{🌸} \times \text{🌸} = 36$$

$$\text{🌸} \times \text{🍃} = 36$$

Common Error & Correction:

Error	Correction
☹ $8 \times 0 = 8$	☺ $8 \times 0 = 0$

Tricks & Shortcuts:

Table of 3

Step 1: Draw lines similar to the game of tic tac toe as shown in the following figure.

Step 2: Write numbers from 1 to 9 starting at the lower left corner.

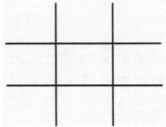

Step 3: Write three zeros in front of the first row, three ones in front of the second row and three twos in front of the third row.

0 0 0 →	3	6	9
1 1 1	2	5	8
2 2 2	1 (Start here)	4	7

Step 4: Insert zeros in front of the first row numbers, one in front of the second row numbers and two in front of the third row numbers.

03	06	09
12	15	18
21	24	27

The table of three is ready. Just read it from left to right in order. It's easy to remember: 3 × 10 = 30.

Finger Calculator

The following trick works for any multiplication of 6, 7, 8, 9 and 10 (Not for 2, 3, 4, 5). For example, it will work for 6 × 8, but not for 6 × 4.

Step 1: Number the fingers as shown in the figure below. You may write the numbers on your finger tip with a sketch pen.

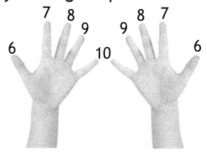

Learning Mathematics - The Fun Way

Step 2: Suppose you want to find 7 × 6, then bend the fingers upto seven on the left hand and upto six on the right hand.

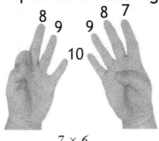

7 × 6

Step 3: Each bend finger is a multiple of ten. In this case, we have three bend fingers (two on the left hand and one on the right), hence, we get 30.

Step 4: Count the number of standing fingers on the left hand. In this case we have three standing fingers.

Left hand has three standing fingers

Step 5: Count the number of standing fingers on the right hand. In this example, we have four standing fingers.

Right hand has four standing fingers

Step 6: Multiply the standing fingers, i.e., 3 × 4 = 12

Step 7: Add the numbers obtained in step 3 and step 6.

30 + 12 = 42. Thus, 7 × 6 = 42.

Multiplication

WORKSHEET

Q1. **Look at the pictures and write the addition sentence and multiplication sentence in each part.**

a)

Addition Sentence   Multiplication Sentence

b)

Addition Sentence   Multiplication Sentence

Learning Mathematics – The Fun Way

Q2. Multiply the number in the centre with the number in the petal. For example, in the first flower, do 3 × 2, 3 × 5, 3 × 9, etc.

Q3. a) Look at the pattern of numbers which are coloured and complete the table.

1	2	3	4	5	6	7	8	9	10
11	12	13	14	15	16	17	18	19	20
21	22	23	24	25	26	27	28	29	30
31	32	33	34	35	36	37	38	39	40
41	42	43	44	45	46	47	48	49	50

b) Look at the number pattern and complete it.

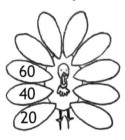

Multiplication

Q4. Multiplication Crossword:

1			2		5
		4			
	3				
6				5	

Across:

    1. 11 × 7

    2. 24 × 14

    3. 50 × 10

    4. 5 × 4

    5. 99 × 1

    6. 15 × 21

Down:

    1. 36 × 2

    2. 12 × 25

    3. 17 × 3

    4. 41 × 5

    5. 16 × 4

Learning Mathematics - The Fun Way

Q5. **a)** Ruhi, Jaswinder, Shahnaaz, Jessica and Rahul love to play with balls. They had five balls each. How many balls do they have in all?

Solution by:

   i. Drawing a picture

   ii. Adding   _____

  iii. Skip counting   _____

   iv. Multiplying   _____

b) A shopkeeper had nine boxes. He kept eight pencils in each box. How many pencils did he have in his shop?

Solution by:

   i. Drawing a picture

   ii. Adding   _____

  iii. Skip counting   _____

   iv. Multiplying   _____

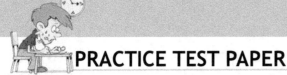

PRACTICE TEST PAPER

Time: 45 minutes   Maximum marks: 10

Q1. Complete the number pattern: **(2 marks)**

    a. 30, 60, 90, ____ , ____, ____.

    b. 125, 100, ____, ____, ____.

Q2. Fill in the blanks: **(2 marks)**

    a. $98 \times 0 =$ ____

    b. $40 \times 10 =$ ____

Q3. Find the product: **(2 marks)**

    a.   26   b. $87 \times 9 =$ _____

        $\times$ 12

        ————

Q4. There are four flower vases in a room. Each vase has eight flowers. How many flowers are there in all?   **(2 marks)**

Solution by:

    i. Drawing a picture

   ii. Adding   _____

  iii. Skip counting   _____

  iv. Multiplying   _____

Learning Mathematics - The Fun Way

DIVISION

Learning Objective:

To relate division with multiplication.

Number of Players:

Two to four

Materials Required:

Sketch pens, paper, thin cardboard, a pair of scissors, glue

Procedure:

9	5	7	63		3	8	6	24
8	6	7	56		5	9	6	30
6	9	4	54		2	7	3	14
4	7	8	32		6	4	8	48
8	3	9	72		9	7	4	36
6	2	7	42		4	1	5	20

Division Game Cards

☞ Each player makes division cards as shown above using paper, cardboard, glue, a pair of scissors and sketch pen.

☞ Each player takes a paper and pen before starting the game.

☞ The division cards are put face down (One number in each card is not related to the other three numbers).

☞ The players turn one card at a time by turn.

☞ After a card is turned up, the players have to write the division and multiplication facts.

☞ The player who writes all the facts correct, scores two points.

☞ The game continues till all the cards are turned.

☞ The player with the highest total score wins the game.

Example:

✳ Suppose the first player turns the following card:

| 8 | 3 | 9 | 72 |

✳ The player who writes the following facts first, scores two points.

$72 \div 9 = 8$

$72 \div 8 = 9$

$8 \times 9 = 72$

$9 \times 8 = 72$

✳ The game continues till all the cards are turned.

✳ The player with the highest total wins the game.

Skill Developed:

Multiplication and division facts.

Activity

Learning Objective:

To understand the concept of division by equal grouping.

Materials Required:

Twenty marbles or beads or buttons, five bowls or glasses

Procedure:

- ☞ Take four bowls (or glasses) and twenty marbles (or buttons).
- ☞ Put the marbles (or buttons), one at a time in each bowl.
- ☞ Keep distributing the marbles, till all the marbles are over.
- ☞ Count the number of marbles in each bowl and record your observation.
- ☞ Repeat the activity with two bowls.
- ☞ Fill in the observation table.

Example:

- ✶ Take five glasses and twenty marbles.
- ✶ Put the marbles, one at a time in each glass till they are over.
- ✶ Count the number of marbles in each glass.

Observation:

Number of bowls/glasses	Number of marbles	Number of marbles in each bowl/glass	Division fact
5	20	4	20 ÷ 5 = 4

Result:

Many division facts can be obtained by equal distribution of the same number of items in different number of objects.

Puzzle

The adjacent clock fell from Geeta's hand and broke into four pieces. To her surprise, she found that the clock broke in such a manner that the sum of two parts was twenty and the sum of other two parts was nineteen. How did the clock break?

 Common Error & Correction:

Error: **Correction:**

☹ 5 ÷ 0 = 0   ☺ 5 ÷ 0 = Not defined

Tricks & Shortcuts:

The various division related terms are shown below.

```
                    4  ←——— Quotient
Divisor ——→ 5 | 23    ←——— Dividend
              |—20
                    3  ←——— Remainder
```

Easy Division of a two-digit number (up to 90) by 9 (Without remainder)

When a two-digit number is divided by 9, the quotient is one more than the digit at the tens place. For example,

In 63 ÷ 9, the quotient is 7 (6 + 1, the digit at the tens place is 6).

WORKSHEET

Q1. Fill in the blanks by looking at the pictures. The lollipops are shared equally by two children and the mangoes are distributed equally on the plates.

a)

Number of lollipops = _____

Number of children = _____

Number of lollipops each child gets = _____
(Write the division fact).

b)

Number of mangoes = _____

Number of plates = _____

Number of mangoes on each plate = _____
(Write the division fact).

Q2. a) Count the number of fishes and distribute them equally into five bowls (by drawing fishes in the bowls). Also write the division fact.

Division fact = _____

Learning Mathematics - The Fun Way

b) Count the number of flowers and distribute them equally into six vases (by drawing flowers in the vases). Also write the division fact.

Division fact = _____

Q3. Solve the division question in each branch and write the divisor, dividend and quotient on the leaves.

a)

Divisor

Dividend

Quotient

Eighty one divided by nine

b)

Divisor

Dividend

Quotient

One hundred divided by ten

c)

Forty nine divided by seven

d)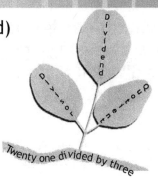

Twenty one divided by three

Q4. The following are the division fact family houses. Use the numbers on the roof to complete the fact families.

a)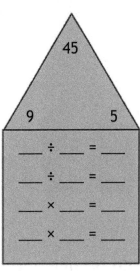

45

9   5

___ ÷ ___ = ___

___ ÷ ___ = ___

___ × ___ = ___

___ × ___ = ___

b)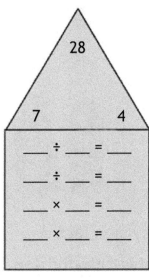

28

7   4

___ ÷ ___ = ___

___ ÷ ___ = ___

___ × ___ = ___

___ × ___ = ___

c)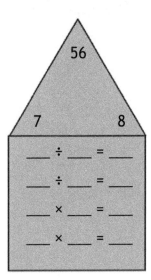

56

7   8

___ ÷ ___ = ___

___ ÷ ___ = ___

___ × ___ = ___

___ × ___ = ___

d)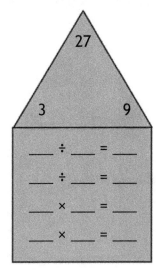

27

3   9

___ ÷ ___ = ___

___ ÷ ___ = ___

___ × ___ = ___

___ × ___ = ___

Learning Mathematics - The Fun Way

PRACTICE TEST PAPER

Time: 45 minutes   Maximum marks: 10

Q1. Two friends share eighteen toffees equally.   (3 marks)

    a. How many toffees will each friend get?_____

    b. __ ÷ __ = _____

    c. If one more friend joins the group and they share equally, then how many toffees will each friend get? _____

Q2. Write the corresponding multiplication facts for each division fact: (4 marks)

Division fact	Multiplication fact
a. 50 ÷ 5 = 10	_____ , _____
b. 40 ÷ 5 = 8	_____ , _____

Q3. Fill in the blanks: (3 marks)

    a. 18 ÷ __ = 2

    b. 16 ÷ __ = 8

    c. __ ÷ 10 = 9

MIRROR HALVES

Game

Learning Objective:

To use a line to divide a shape into two similar halves.

Number of Players:

Two to four

Materials Required:

A pair of scissors, ruler, pencil, paper

Procedure:

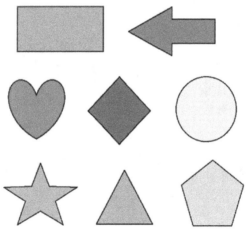

Shape Cards 1

☞ Cut the above shaped cards from the back of the book.

☞ Each player arranges the cards in the same order.

☞ The cards are put face down and turned one by one.

☞ Once the card is turned, every player has to draw as many lines possible to divide the shape into two similar halves.

☞ For each correct line, the player scores one point.

☞ The game continues till all the cards are turned.

Learning Mathematics - The Fun Way

☞ The player with the highest total wins the game.

Example:

✶ Suppose the first player turns the following card:

✶ The player who draws the following lines scores five points.

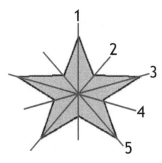

✶ The game continues till all the cards are turned up.
✶ The player with the highest total wins the game.

Skill developed:

Concept of mirror half.

Learning Objective:

To develop the skill of understanding mirror halves.

Materials Required:

A pair of scissors.

Procedure:

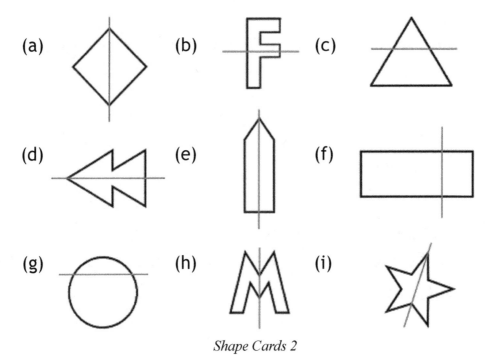

Shape Cards 2

☞ Cut out the above shapes from the back of the book.

☞ Fold along the red line.

☞ Note whether the red line divides each shape into two identical mirror halves.

☞ Record your observation.

Learning Mathematics - The Fun Way

Example:

Yes the red line divides this shape into two identical mirror halves.

Observation:

Shape	Identical mirror halves (Yes/No)
(a)	Yes

Result:

All shapes cannot be divided into two identical mirror halves.

What title of respect was given to Mohandas Karamchand Gandhi?
Draw the mirror image of the following to find the answer.

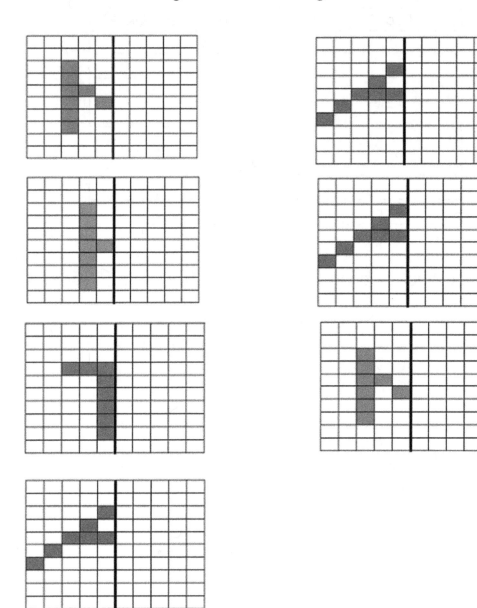

Learning Mathematics - The Fun Way

Common Error & Correction:

Error: Correction:

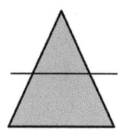   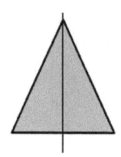

☹ Mirror half of a Triangle   ☺ Mirror half of a Triangle

 Tricks & Shortcuts:

If the sides of a shape are equal in length, then the number of lines of mirror halves is equal to the number of sides of the shape.

Shape	Number of Lines of Mirror Halves
	3
	4
	5

WORKSHEET

Q1. Does the dotted line in each picture, divide it into two similar mirror halves? Write 'Yes' or 'No'.

(a)

(b)

(c)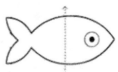

_____   _____   _____

(d)

(e)

(f)

_____   _____   _____

Q2. Draw a dotted line to divide each of the following pictures into two similar halves.

Q3. Draw the mirror half of the following pictures:

Learning Mathematics - The Fun Way

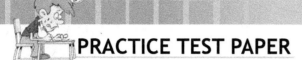

PRACTICE TEST PAPER

Time: 45 minutes   Maximum marks: 10

Q1. Does the given line in each picture divide it into two similar mirror halves? Write 'Yes' or 'No'.   (2 Marks)

_____   _____

Q2. Draw a dotted line to divide each of the following pictures into two similar halves.   (2 Marks)

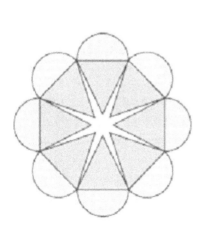

Q3. What is the name of this dinosaur?

Find out by completing the mirror halves of the following letters. **(2 Marks)**

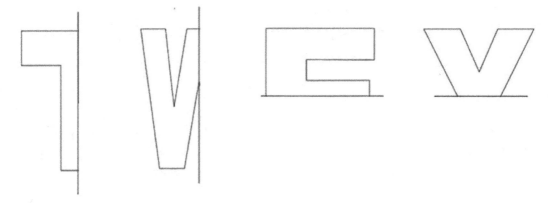

Q4. Colour the circles in such a way that the dotted line divides each board into two similar halves. **(4 Marks)**

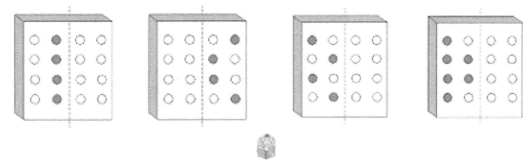

Learning Mathematics - The Fun Way

SHAPES AND DESIGNS Game

Learning Objective:

To explore different shapes using given geometrical shapes of the tangram set.

Number of Players:

Two

Materials Required:

A pair of scissors, thin cardboard, glue

Procedure:

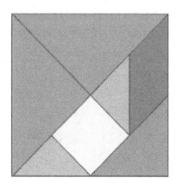

Tangram Set

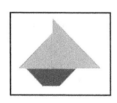   

Tangram Designs

- ☞ Cut out the tangram set from the back of the book.
- ☞ Paste them on a thin cardboard.
- ☞ Each player should have one set of tangram shapes.
- ☞ Cut out the tangram design shape cards.

- ☞ Put them face down.
- ☞ Each player turns a design card.
- ☞ The player who can make the design first by using all the tangram shapes scores five points.
- ☞ The game continues till all the designs are made.
- ☞ The player with the highest total score wins the game.

Example:

- ✳ Suppose the first player turns the following boat design:

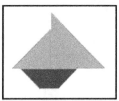

- ✳ The player with the following arrangement scores five points.

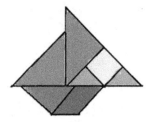

- ✳ The game continues till all the designs are made using tangram shape pieces.
- ✳ The player with the highest total score wins the game.

Skill developed:

Develops spatial understanding.

Learning Mathematics - The Fun Way

Activity

Learning Objective:

To identify the edges and corners of different shapes.

Materials Required:

Yellow and black square paper, glue, a pair of scissors

Procedure:

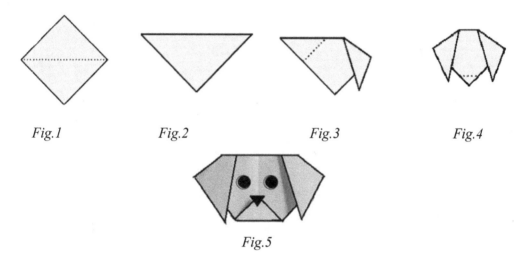

Fig.1   Fig.2   Fig.3   Fig.4

Fig.5

- ☞ Take a yellow square paper as shown in figure 1.
- ☞ Fold it along the dotted lines.
- ☞ Fold from the two corners of the triangle as shown in figure 3.
- ☞ Draw a dotted line as shown in figure 4 and fold it.
- ☞ Cut two small circles (you may use a twenty-five paise coin) and a triangle to make the eyes and the nose of the dog.
- ☞ Note the number of sides and corners in each figure and fill in the observation table.

Observation:

Figure number	Number of sides	Number of corners
Fig. 1	4	4

Result:

The number of sides and corners varies from shape to shape.

Learning Mathematics - The Fun Way

Puzzle

Which shape can you make by using the following eight pieces?

_____

 Common Error & Correction:

Error:	Correction:
Number of edges of a sphere = 1	Number of edges of a sphere = 0

Tricks & Shortcuts:

Learn the following poems to remember the number of edges, corners and faces of cube, cuboid, sphere and cone.

Cube and **cuboid** took part in races,

Each of them had **six** clean **faces**.

Both of them reached very late,

The number of **corners** each has **eight**.

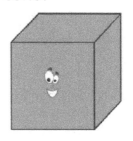

The running ground had green tall hedges,

The two 3D shapes have **twelve** straight **edges**.

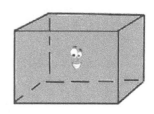

Mr. **Sphere** is a very good learner

He has **zero edge** and **zero corner**.

He loves to eat butter and bun,

The number of **face** he has **one**.

Mrs. **Cone** loves the zoo,

The number of **faces** she has **two**.

At the zoo she had some fun,

Edges and **corner** she has **one**.

WORKSHEET

Q1. Write 'C' if the figure is made of only curved lines, 'S' if the figure is made of only straight lines and 'C & S', if it is made of both curved lines and straight lines.

(a) _____

(b) _____

(c)  _____

(d) _____

(e)  _____

(f) _____

(g) _____

(h) _____

Q2. Fill in the following tables:

a)

Shape	Shape name	Number of sides	Number of corners
◼			
▲			

b)

Shape/Shape name	Number of faces	Number of edges	Number of corners
_____			
_____			
_____			
_____			

Learning Mathematics - The Fun Way

Q3. Complete the following tiling patterns:

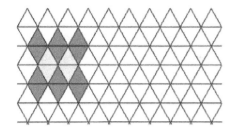 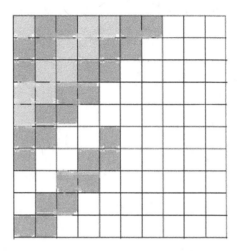

Q4. Draw the following *rangoli* pattern on a dot grid.

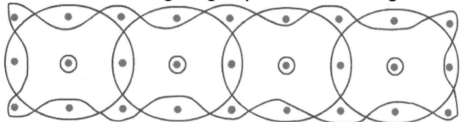

Q5. Write whether the following pictures show the top view or the side view.

(a)

_____

(b)

_____

(c)

_____

(d)

_____

Q6. See the map and follow the instructions given below. Draw your path.

Learning Mathematics - The Fun Way

a) Start at the vegetable shop and turn left.

b) Go to the flower bed and colour the smallest flower.

c) Now go to the restaurant and circle the fork and spoon in front of the door.

d) Next go to the school and draw a star on the highest dome of the building.

e) Follow the path to the library and colour the trees behind it.

f) Finally, go towards the playground, cross the bridge and colour the nearest tree.

g) Move right and circle the bench.

PRACTICE TEST PAPER

Time: 45 minutes   Maximum marks: 10

Q1. Complete the following table: **(4 Marks)**

Shape	Number of sides	Number of corners
Rectangle		
Pentagon		
Oval		
Cube		

Q2. a. Draw a figure using two straight lines and two curved lines.   **(1 Mark)**

b. Which of the following objects will always have corners?

Objects made of straight edges or objects made of curved edges.   **(1 Mark)**

_____

Learning Mathematics - The Fun Way

Q3. Colour the tiles which can cover the floor completely without leaving any gap between them. **(2 Marks)**

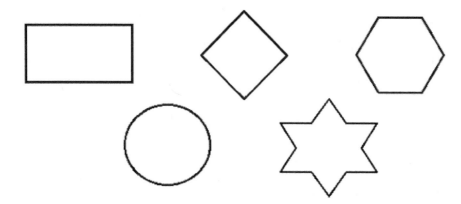

Q4. **a.** Rearrange two matchsticks to get four similar squares.

(2 Marks)

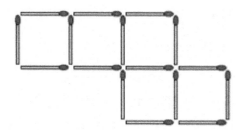

b. Rearrange three matchsticks to get three similar squares.

PATTERNS

Game

Learning Objective:

To arrange the names of different things in an alphabetical order.

Number of Players:

Two to four

Materials Required:

Paper, pencil, thin cardboard, a pair of scissors

Procedure:

☞ Make cards with names of five states, five fruits, five vegetables, five friends, etc.

☞ Make as many cards as you want.

☞ Put them face down at the centre.

☞ Turn the cards at the centre one at a time.

☞ Each player takes a paper and pencil and writes the names in alphabetical order.

☞ The first player to get it correct scores four points.

☞ The player with the highest total in the end is the winner.

Example:

Kerala	Tripura
Maharashtra	Bihar
Himachal Pradesh	

If the above card is turned, the player who writes Bihar, Himachal Pradesh, Kerala, Maharashtra, Tripura, first scores four points.

Skill developed:

Ordering of alphabets.

Learning Mathematics - The Fun Way

Activity

Learning Objective:

To develop the skill of growing patterns.

Materials Required:

Drawing sheet, crayons, pencil, ruler and a pair of scissors.

Procedure:

- ☞ Draw a square. Use a ruler to connect one pair of opposite corners.
- ☞ Colour the two triangles using different colours.
- ☞ Draw 16 similar squares.
- ☞ Use the squares to make the following patterns:

 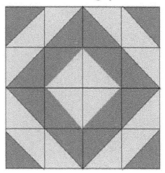

- ☞ Record the observation.

Example:

Draw the following shapes on the drawing sheet.

Draw 16 such squares and make the given patterns.

Observation:

* ✯   When we join the opposite corners of the square, we get _____ triangles.

* ✯   Using 16 similar squares divided into triangles, we can make _____ (same/different) patterns.

Result:

Using similar squares divided into triangles, different patterns can be obtained.

Learning Mathematics - The Fun Way

Puzzle

There are some words which remain the same when you read it backwards. E.g. Mom, dad, noon, etc. Name a body part which follows the same pattern.

_____

 Common Error & Correction:

Error:

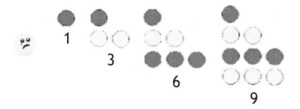

Correction:

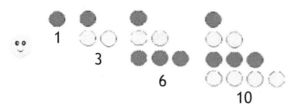

 Tricks & Shortcuts:

In repeating patterns, no new things are added, whereas in growing pattern, something new is added. Look for the rule carefully.

WORKSHEET

Q1. Complete the following design patterns:

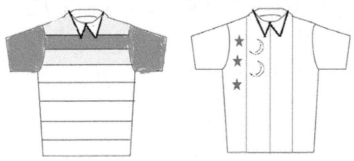

Q2. Find the rule and complete the following growing patterns:

a)

b)

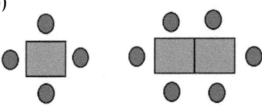

Q3. Find the rule and complete the following number patterns:

    a) 9, 18, _____, 36, _____, _____.

    b) 11, 22, 33, _____, _____, _____.

    c) 20, 18, 16, _____, _____, _____.

    d) 25A, _____, 75C, 100D, _____, _____.

Q4. Complete the following pattern with odd numbers:

    a) 11, 13, 15, _____, _____, _____.

Learning Mathematics - The Fun Way

b) In a computer, how many different digits are used to store the information? Colour only the odd numbers in the following table to find the answer.

1	13	35	69	47	39	71	51	3
49	17	2	14	50	26	10	15	31
79	43	91	61	83	21	38	87	59
41	25	37	15	95	3	42	45	65
97	9	8	18	46	34	16	11	75
33	63	20	19	53	41	73	85	93
13	55	32	89	99	81	23	57	27
83	69	24	28	40	4	30	53	81
5	71	49	37	67	99	23	79	7

Q5. The following people have their cell phone numbers according to their names. Use the following cell phone keypad to find their numbers.

987JAYANTI   _____   (J = 5, A = 2, etc.)

981SAURABH   _____

983JAIDEEP   _____

805SHIKSHA   _____

PRACTICE TEST PAPER

Time: 45 minutes   Maximum marks: 10

Q1. Complete the following number patterns: (3 Marks)

    a. 12, 14, 16, ___, ___. These numbers are called ___ numbers.

    b. 63, 56, 49, ___, ___.

Q2. Find the rule and complete the following patterns: (2 Marks)

    a. A, AB, ABC, _____, _____.

    b. Apple 1, _____, Cherry 3, Date 4.

Q3. Find the rule and draw the next two pictures in each part in the following: (2 Mark)

    a.

                                                      _____ _____

    b.

                                                      _____ _____

Q4. Answer the following questions: (3 Marks)

    a. Which of the following fruits will appear first in a dictionary?

    b. Which of the following names will appear first in a telephone directory?

       Mr. Khan, Mr. Fernandes, Mr. Mehta, Mr. Singh

    c. If A = 1, B = 2, C = 3 and so on, which number will P represent?

Learning Mathematics - The Fun Way

LENGTH

Learning Objective:

To match the stamps with the envelope based on the length of the envelope.

Number of Players:

Two

Materials Required:

A pair of scissors, four paper clips, ruler.

Procedure:

   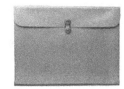

Envelopes

Stamps

☞ The players cut out the envelopes and the stamps from the back of the book.

☞ The game starts after both the players cut out the envelopes and the stamps.

☞ Each player measures the envelopes, checks the chart and puts the stamps on the correct envelope with the paper clip.

Size of the envelope	Value of the stamp (in rupees)
Less than 5 cm	5
5-10 cm	10
11-15 cm	15
16-20 cm	20

☞ The player who can first clip all the four stamps with the envelopes in the correct order is the winner.

Example:

✶ If the length of the envelope is 12 cm, clip the 15-rupee stamp to it.

Skill Developed:

Measuring the length using a ruler. Decision making on the correct value of the stamp.

Learning Mathematics - The Fun Way

Activity

Learning Objective:

☞ To develop the skill of measuring length.

☞ To collect and interpret data.

Materials Required:

Pencil, brown paper or old newspaper, measuring tape.

Procedure:

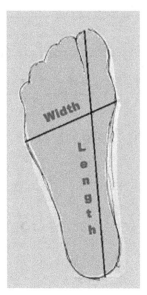

☞ Draw the outline of your feet on a brown paper (or an old newspaper).

☞ Measure the distance from the tip of the toe to the end of the heel. This is the length of the feet.

☞ Now measure the width of the feet.

☞ Do the same for your whole family and record your observation.

☞ Find out whose feet is longest in size and whose feet are widest in size.

Example:

Mother's length of the feet = 21cm

Mother's width of the feet = 7cm

Observation:

Name or relationship	Length of the feet	Width of the feet
Father		
Mother		
Myself		
Brother/Sister		

In my family, _____ has the longest feet and _____ has the widest feet.

Result:

The length and width of the feet varies from person to person.

 Puzzle

Suresh said, "My height is two centimetres more than a metre rod."

Mahesh said, "My height is one hundred and three centimetres."

Who is taller?

 Common Error & Correction:

Error	Correction
1m 5cm = 150cm	1m 5cm = 105 cm

 Tricks & Shortcuts:

The word, "kilo" stands for 1000. Hence, kilometre is used for measuring long distances (for example, distance between two places like two cities, two countries, etc.)

WORKSHEET

Q1. How many buttons will cover the length of the pencil?

_____

Q2. Fill in the blanks:

a) _____, _____ and _____ are the commonly used units of length.

b) The ruler in a geometry box measures up to a length of _____ centimetre.

c) A metre rod can measure up to a length of _____ centimetre.

Q3. Write 'more than one metre' or 'less than one metre' in front of each of the following:

a) Length of my school bag is _____

b) My class teacher's height is _____

c) The cloth required to make my shirt : _____

d) The width of my school table : _____

Q4. First guess the length of each of the following and then verify it by actual measurement.

Object	My guess	Actual measurement
Width of my Mathematics book		
My mother's saree		
My waist		
My friend's height		

Learning Mathematics – The Fun Way

Q5. Look at the following picture and answer the questions given below.

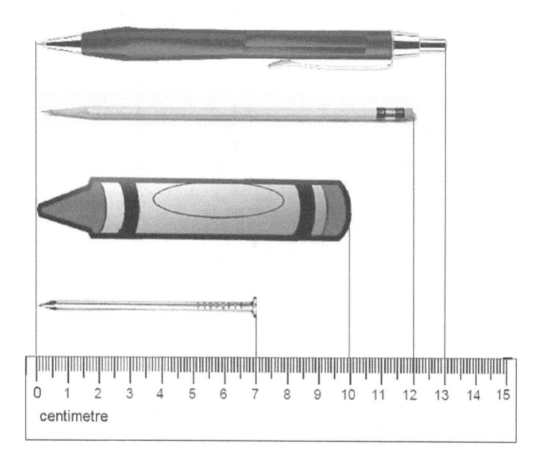

a) What is the length of the crayon? _____

b) What is the length of the longest object? _____

c) What is the length of the shortest object? _____

d) What is the difference in length of the pencil and the crayon? _____

Q6. Look at the following map of Kolkata, study the distances and answer the questions given below:

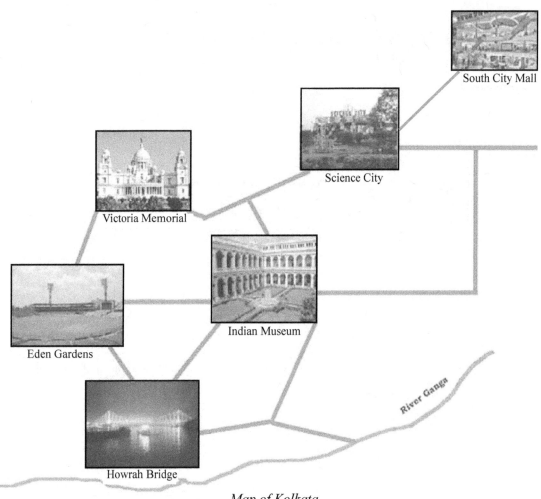

Map of Kolkata

- Distance between Howrah Bridge and Eden Gardens = 3 Km
- Distance between Howrah Bridge and Indian Museum = 4 Km
- Distance between Victoria Memorial and Science City = 7 Km
- Distance between Science City and South City Mall = 9 Km
- Distance between Indian Museum and Science City = 6 Km

Learning Mathematics - The Fun Way

a) Which place is closer to Howrah Bridge: Eden Gardens or Indian Museum? _____

b) Which is farther from Indian Museum: Victoria Memorial or Science City? _____

c) What is the distance between Victoria Memorial and South City Mall? _____

d) Name the river over which the Howrah Bridge lies? _____

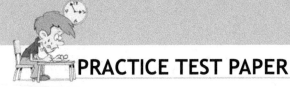

PRACTICE TEST PAPER

Time: 45 minutes   Maximum marks: 10

Q1. Write the units used for measuring the following items. Write 'Km' for Kilometre, 'm' for metre and 'cm' for centimetre.   **(3 Marks)**

a. Distance between Delhi and Mumbai: _____

b. Height of a newborn baby: _____

c. Length of a dining table: _____

Q2. Name a profession in which the following measuring device is used.   **(1 Mark)**

_____

Q3. Measure the length of your foot from the tip of toe to the end of heel. How short is it from the standard foot unit?

(1 foot = 30 cm)   **(2 Marks)**

_____ , _____

Q4. Shruti is going to her grandmother's house. She can take three roads: A, B or C. Which is the shortest?   **(2 Marks)**

_____

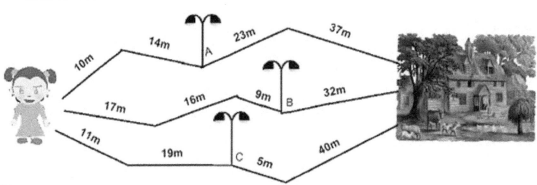

Learning Mathematics - The Fun Way

Q5. Measure the length of the following straw. (1 Mark)

_____

Q6. Look at the map of India and tell which city is closer to New Delhi – Bangalore (Bengaluru) or Bhopal? (1 Mark)

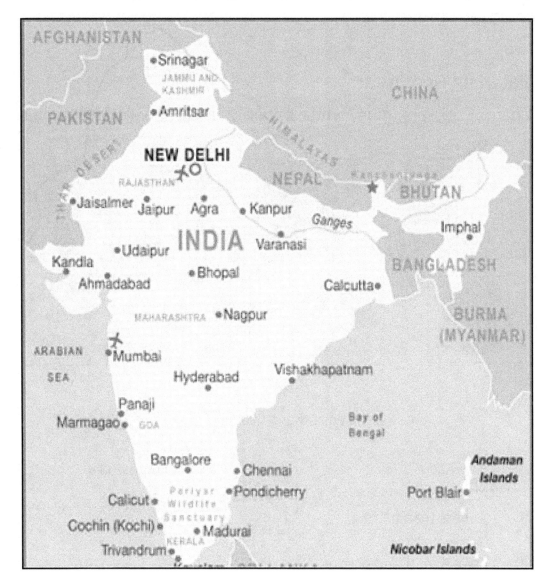

WEIGHT

Game

Learning Objective:

To estimate weights of objects.

Number of Players:

Two

Materials Required:

A pair of scissors, glue, white paper, thin cardboard, sketch pens

Procedure:

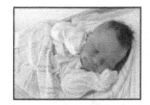  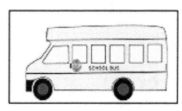

Object Cards

More than 1000 kg

About 100 kg

About 10 kg

Less than half kg

About 3 kg

Between half and one kg

Lighter than football

Weight Game Cards

Learning Mathematics - The Fun Way

- ☞ Cut out the above object cards from the back of the book.
- ☞ Make the above weight cards by writing the weights on a piece of paper and cutting and pasting on a thin cardboard.
- ☞ Each player makes a set of weight game cards.
- ☞ The object cards are kept face down at the centre.
- ☞ Each player turns a card by turn.
- ☞ The player has to match the weight of the object.
- ☞ The player putting the correct weight card first next to the object card scores two points.
- ☞ The game continues for seven rounds.
- ☞ The player with the highest total score wins the game.

Example:

* Suppose the first player turns the following picture:

* The player who first puts the following card next to it scores two points.

About 10 kg

* The game continues for seven rounds.
* The player with the highest total score wins the game.

Skill developed:

Estimation of weights, matching weights with objects.

Learning Objective:

To develop the understanding of weight in terms of non-standard units.

Materials Required:

Paper cup, thread, rubber bands, paper clip, ruler, thick book, table, chart paper, pencil, eraser, sharpener and a small potato.

Procedure:

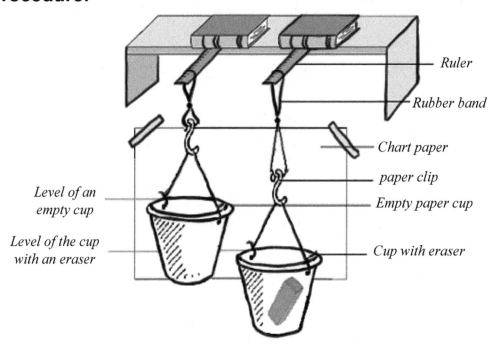

☞ Tie two rubber bands and put one of them into the ruler as shown in the above diagram.

☞ Put the ruler under a thick book kept on the corner of the table.

☞ Twist a paper clip and put one end into the rubber band and the other end into the thread of the paper cup.

Learning Mathematics - The Fun Way

- ☞ Fix a chart paper on the wall behind the paper cup.
- ☞ Mark the level of the empty paper cup on the chart paper.
- ☞ Now put different objects like sharpener, eraser and potato one by one into the paper cup and mark their levels on the chart paper with a pencil.
- ☞ Observe with which object the rubber band is stretched maximum.

Example:

The plastic sharpener is lighter than the eraser.

Observation:

a) The _____ is heavier than the _____.

b) The _____ is lightest.

c) The _____ is heaviest.

Result:

The heavier the object, the greater will be the stretch in the rubber band.

Which is heavier: one kilogram of iron or one kilogram of cotton?

_____

 Common Error & Correction:

Error	Correction
The bigger the size of an object, the heavier it is.	The weight of an object does not depend on its size.

For example, in the given picture, the balloon has a bigger size than the brick, but the brick is heavier than the balloon.

 Tricks & Shortcuts:

The word, "kilo" stands for 1000. Hence, kilogram means 1000 grams.

WORKSHEET

Q1. Arrange the following objects in the increasing order of their weights by writing numbers, 1 to 4 below them.

a)

_____   _____   _____   _____

b)

_____   _____   _____   _____

Q2. What is the weight of the toy car in terms of oranges?

_____

Q3. The weighing balance shows 500 grams, 1 kilogram, 1.5 (one and half) kilogram and 2 kilograms. Look at the pictures and answer the questions given below.

Pineapple   *Apple*   *Apricot*

a) Which fruit is the heaviest? _____

b) Which fruit is the lightest? _____

c) How much does apple weigh? _____

d) How much does apricot weigh? _____

Q4. Draw the missing balls on the right hand side of the balance to make both sides equal.

Q5. Do you weigh more than Raghav or less than Raghav? How much more or how much less?

_____

I am small in front of an elephant. My weight is only 30 Kg.

Q6. Name the following items from daily life:

a) Two items which weigh less than half kilogram: _____

b) Two items which weigh more than one kilogram: _____

Q7. Fill in the blanks by choosing the correct number:

a) 1 Kilogram = _____ grams (500 / 1000)

b) A kitchen gas cylinder weighs about _____ kilograms. (2/14)

c) A paper clip weighs about _____ gram. (1 / 100)

Learning Mathematics - The Fun Way

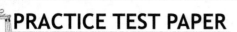

PRACTICE TEST PAPER

Time: 45 minutes   Maximum marks: 10

Q1. **Write the units used for measuring the following items. Write 'g' for gram and 'Kg' for kilogram.** **(4 Marks)**

    a. Weight of a coin   _____

    b. Weight of a tiger   _____

    c. Weight of a pencil   _____

    d. Weight of twelve apples   _____

Q2. **Which of the following fruits weigh more than one kilogram and which ones weigh less than one kilogram?** **(2 Marks)**

_____   _____

_____   _____

Weight

Q3. Which of the following will have a bigger bag:

One kilogram of puffed rice (known as *murmura* in Hindi) or one kilogram of wheat? (1 Mark)

_____

Q4. Susan, Salma and Swati want to participate in a gymastic competition. The minimum body weight has to be 30 kilograms for participation. Salma weighs five kilograms more than Susan. Swati weighs less than both Susan and Salma. Susan weighs 25 kilograms. Who can participate in the competition? (1 Mark)

_____

Q5. Where can you see the following weighing scales: (2 Marks)

a.   b.

_____   _____

Learning Mathematics - The Fun Way

CAPACITY

Learning Objective:

To measure the capacity of different containers in terms of non standard units.

Number of Players:

Two

Materials Required:

A pair of scissors, sketch pens

Procedure:

Capacity Game Board

About 4	About 10	About 1
About 11	About 2	About 100
About 30	About 6	About 7

Capacity Game Cards

☞ Cut out the above capacity game cards from the back of the book.

☞ Each player should have his/her own set of nine cards with his/her name initials marked.

☞ The game is played like tic-tac-toe (knots and crosses).

☞ The players by turn put a matching card on the game board.

☞ If the player matches it wrong, then he/she loses one turn.

☞ The first player with three cards in a row, column or corner to corner (diagonal) wins the game.

Example:

✻ If Pushpa and Shaheen are playing the game, then Pushpa should mark her cards with "P" and Shaheen should mark them with "S".

✻ The following card should be placed on top of the bucket.

✻ The game continues till any player gets three cards in a row, column or diagonal.

✻ If no one gets it, then the game ends in a draw.

P
About 11

Skill developed:

Measurement of volume in terms of non-standard units.

Learning Objective:
To compare the volume of water in different containers.

Materials Required:
Containers of different shapes and sizes.

Procedure:
- ☞ Take two glasses of the same shape and size.
- ☞ Pour water in them up to the same level.
- ☞ Now take one of the glasses and pour the water into another glass of different size.
- ☞ Note if the water level rises and record your observation.
- ☞ Repeat the activity by pouring water into different containers.

Example:
- ✷ Pour water up to the same level in the following two glasses.

- ✷ Now take one of the glasses and pour the water into a narrow tall glass.

Observation:

In the narrow glass, the level of water is _____ (higher/lower).

Result:
We can compare the volume of the liquid by the height of the liquid, if the containers are of the same shape and size.

Puzzle

Mrs. Iyer had two bowls measuring four litres and three litres. She wanted to measure two litres of water using these two bowls. Can you help her out?

_____

_____

Common Error & Correction

Error

😕 If the level of a liquid is higher in a container, it has a greater volume.

Correction

😊 The volume of the liquid can be compared by seeing the level if the two containers are of the same shape and size.

Tricks & Shortcuts

The liquids always take the shape of the container in which they are poured. Hence, we cannot compare the volume of liquids just by seeing the levels.

WORKSHEET

Q1. Arrange the following objects in the increasing order of their capacities by writing numbers, 1 to 4 below them.

a) _____ _____ _____ _____

b) _____ _____ _____ _____

Q2. A scientist poured different liquid chemicals in various containers. Look at the containers and answer the questions given below.

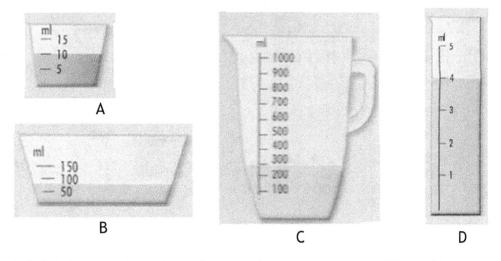

a) Which container has the maximum amount of liquid? _____

b) Which container has the minimum amount of liquid? _____

c) What is the difference between container A and D? _____

Capacity

d) How much chemical should he pour in vessel C to make it 400 ml? _____

Q3. Rishabh and Robert each bought a fish bowl from a pet shop. The shopkeeper told them that the fishes can swim comfortably in half litre (or 500 ml) of water. Rishabh poured one glass of water in his bowl and Robert poured two glasses of water in his bowl. Whose fish is more comfortable? (One glass of water is about 250 ml)

_____

Q4. See the water mark in the big bottle.

Ten small bottles of water are poured into the big bottle. How many more bottles can be poured to fill the big bottle completely?

_____

Q5. Fill in the blanks by choosing the correct numbers:

a) We should drink about _____ glasses of water everyday. (6 to 8/ 26 to 28)

b) Mother dairy milk packets contain ___ litre or ___ litre of milk. (half or one / one or two)

Time: 45 minutes   Maximum marks: 10

Q1. Write the units used for measuring the following items. Write 'l' for litre and 'ml' for millilitre.   (4 Marks)

a. Petrol in a car   _____

b. Small bottle of soft drink   _____

c. One teaspoon of milk   _____

d. Water in a water tank   _____

Q2. Which of the following containers can hold more than one litre of water and which ones can hold less than one litre of water?   (2 Marks)

a.   b.

_____   _____

c.   d.

_____   _____

Q3. Fill in the blanks by choosing the correct numbers:

                                                                (2 Marks)

a. One teaspoon can hold about _____ ml of water. (5 / 50)

b. One cup can hold about _____ ml of milk. (600 / 250)

Capacity

Q4.

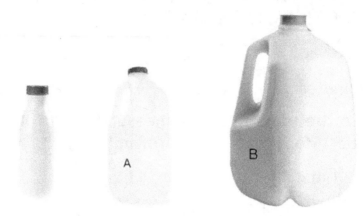

Jug B contains double the quantity of milk as compared to jug A. If jug A contains four bottles of milk, how many bottles of milk are contained in jug B?   (2 Marks)

_____

 **TIME**

Learning Objective:

To develop the skill of reading and arranging time in order.

Number of Players:

Two

Materials Required:

White or coloured paper, thin cardboard, ruler, a pair of scissors, a bangle, glue, sketch pens.

How to play:

☞ Draw clock faces on a paper using a bangle.

☞ Use a ruler to draw the hands.

☞ Draw all the full hours as well as the half hours from 1 to 12. (Some of them are shown above).

☞ Cut the clock faces and paste them on a thin cardboard.

☞ Now you have twenty-four clock cards (12 full hours and 12 half hours).

☞ Shuffle the clock cards and distribute three cards to each player (The cards are not to be shown to the other players. Each player sees his/her own card).

☞ Put the remaining cards face down at the centre.

☞ Each player takes a card from the centre by turn and throws

down one card. (He/she may throw the same card or any other card from his/her hand).

☞ The player has to arrange a set of three cards in his/her hand in order. It can be in order of full hour, half hour or mixed.

☞ The winner is the player who can arrange the ordered set first.

Example:

* Suppose Sukhvinder and Shivani are playing the game.

* Both get three clock cards and the remaining cards are put face down at the centre.

* Suppose Sukhvinder has six o'clock, four o'clock and seven o'clock and Shivani has nine o'clock, one thirty and four thirty.

* Sukhvinder picks a card from the centre and gets seven thirty. He keeps that and throws four o'clock.

* Now Shivani picks a card and gets five o'clock. She keeps that and throws nine o'clock.

* Next Sukhvinder gets six thirty. He keeps that and throws six o'clock.

* He arranges his cards as follows and wins the game.

Skills developed:

Reading half hour and full hour, arranging the clock time in order.

Learning Mathematics – The Fun Way

Learning Objective:

To find out the position of the big hand (minute hand) at full hour and half hour.

Materials Required:

A pair of scissors, crayons, glue, thin cardboard or paper plate, thumb pin.

Procedure:

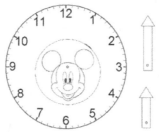

☞ Cut out the clock face and the hands of the clock from the back of the book. Colour them. Paste the clock face on a paper plate or on a thin cardboard.

☞ Fix the two hands at the centre (on the forehead of a mickey mouse) with a thumb pin.

☞ Move the hands to show the following time:

☞ 5 O'Clock, 11 O'Clock, 8 O'Clock, 12 O'Clock, 4:30, 2:30, 7:30

☞ Fill in the observation table and the blanks.

Time

Example:

3 O'Clock

Observation:

Time	Minute hand (big hand) at	Hour hand (small hand) at
3 O'clock	12	3

At full hour, the minute hand (big hand) is always at _____.

At half hour, the minute hand (big hand) is always at _____.

Result:

At full hour, the minute hand is always at 12, whereas at half hour, the minute hand is always at 6.

Learning Mathematics - The Fun Way

Puzzle

The following clock shows the reflection of a clock in a mirror. What is the actual time?

(Hint: In a mirror, the left looks right and the right looks left).

 Common Error & Correction:

Error	Correction
Time = 3 : 30	Time = 2 : 30

 Tricks & Shortcuts:

For half hours, remember the following:

When the hour hand is in the middle, choose the number that is little.

For example, when the hour hand is between two and three, it is two thirty and not three thirty.

WORKSHEET

Q1. List the following events in the table given below:

a) Blinking of eyes, Watching a movie, Cooking kidney beans (rajmah), Change of traffic lights, Sleeping at night, Taking off shirt.

Takes seconds	Takes minutes	Takes hours

b) Growing of a child into an adult, Change of season, Growing plants from the seeds, Construction of a four-storey house, Making of a hand embroidered saree.

Takes days	Takes months	Takes years

Q2. Write the following clock time:

a)

b)
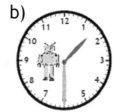

c)
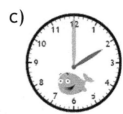

_____   _____   _____

Learning Mathematics - The Fun Way

d) 　　　e)

_____　　　_____

Q3. Draw the hands of the clock showing the time given below each clock.

a)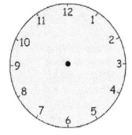

10 O'Clock

b)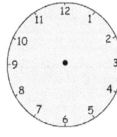

2:30

c)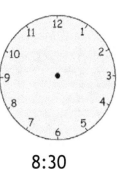

8:30

d)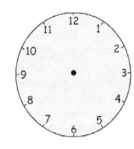

7 O'Clock

Q4. Some activities are described below. Write the time of the activity. Also draw the hands of the clock showing the time.

I wake up in the morning at

_____
Time

My school starts at

_____

My school gets over at

_____

I do my homework at

_____

I go to bed at

_____

Learning Mathematics - The Fun Way

PRACTICE TEST PAPER

Time: 45 minutes   Maximum marks: 10

Q1. Write whether it takes seconds, minutes, hours, days, months or years to do each of the following activity: (3 Marks)

    a. Giving high five   _____

    b. Air travel between Mumbai and New Delhi _____

    c. Movement of the Earth around the Sun _____

Q2. Write the following clock time: (2 Marks)

a.    b.

    _____   _____

c.    d.

    _____   _____

Q3. Answer the following questions: (3 Marks)

    a. At what time do we find both the hands of a clock at the same number? _____

    b. Lata's brother was born in 2010. How old is he now? _____

c. Ramu's sister was born on 1/1/2011. On what date will she be ten years old? _____

Q4. Draw the hands of the clock showing the time given below each clock. (2 Marks)

a.

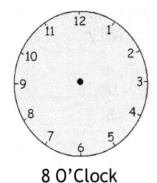

8 O'Clock

b.

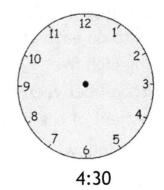

4:30

Learning Mathematics - The Fun Way

CALENDAR

Learning Objective:

To show a mathematical magic based on a calendar month by giving proper sequence of instructions.

Number of Players:

Two

Materials Required:

Calendar, paper and pencil

SUN	MON	TUES	WED	THURS	FRI	SAT
2	31					1
9	3	4	5	6	7	8
16	10	11	12	13	14	15
23	17	18	19	20	21	22
30	24	25	26	27	28	29

Procedure:

☞ Select any month of the calendar.

☞ Ask your friend to choose any four dates, such that they form a square.

☞ Now ask him/her to add the four dates and tell you the sum.

☞ Just by knowing the sum, you can tell the four dates.

Example:

✫ Suppose your friend selects four dates as 4,5,11 and 12.

✫ Sum = 4 + 5 + 11 + 12 = 32

Trick:

Divide the sum by four (32 ÷ 4 = 8) and then subtract four (8 – 4 = 4). This gives the first date. Add 1, 7 and 8 to get the other three dates (4 + 1= 5, 4 + 7 = 11 and 4 + 8 = 12).

Skill developed:

Mental addition, subtraction and division of numbers.

Learning Objective:

To identify a festival by seeing its picture and find out its date and day by reading a calendar.

Materials Required:

Current year's calendar

Procedure:

☞ See the pictures carefully and write the name of the festivals.

☞ Take current year's calendar and find out the date and month of the festival.

☞ Now find out the day on which the festival falls.

☞ Fill in the observation table.

Example:

If we take the calendar of 2012, we will get:

Festival	Name of the festival	Date and Month	Day
	Holi	8th March	Thursday

Observation:

Festival	Name of the festival	Date and Month	Day

Learning Mathematics - The Fun Way

Result:

Different festivals fall on different dates, months and days.

Puzzle

I can't find my birthday in the calendar!

Rishabh was born in a leap year. In a leap year there are 366 days instead of 365. A leap year repeats after four years. Rishabh saw the calendar in a non-leap year and could not find his date of birth. What is his date of birth? (Find the date and the month, not the year).

 ## Common Error & Correction:

Error	Correction
In a leap year, February has 28 days.	In a leap year, February has 29 days.

 ## Tricks & Shortcuts:

Make your fist (as shown in the figure) and start counting on any knuckle (the raised finger joint) with January. The space between the knuckles represents February. Again the raised knuckle is March. Continue counting by knuckle and the space between the knuckles. The knuckle is higher so the months on the knuckle have 31 days. The space between the knuckles is lower. Hence, the months on these spaces have 30 days, except February which has 28 days in an ordinary year and 29 days in a leap year.

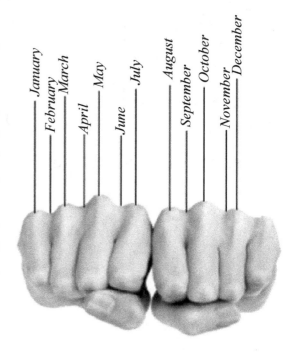

January, February, March, April, May, June, July, August, September, October, November, December

WORKSHEET

Q1. Unscramble the names of the months in the following tree:

a) _____

b) _____

c) _____

d) _____

e) _____

f) _____

g) _____

h) _____

i) _____

j) _____

k) _____

l) _____

Q2. Find the date and month of the following:

a) India's Independence Day _____

b) Gandhi Jayanti _____

c) Children's Day (in India) _____

d) Teachers' Day (in India) _____

Q3. Fill in the dates in the following calendar and answer the questions given below:

July

SUN	MON	TUES	WED	THURS	FRI	SAT
8						
				19		

a) Which date falls on the third Wednesday of this month? _____

b) How many weeks are there in this month? _____

c) Can the second Saturday of any month be on the 15th? _____

d) Which days occur five times in this month? _____

Q4. Answer the following questions:

a) What is your date of birth? _____

b) In which year did you take admission in this school? _____

c) In which year will you be in Class V? _____

d) Name the months which have thirty-one days. _____ _____

Learning Mathematics – The Fun Way

PRACTICE TEST PAPER

Time: 45 minutes   Maximum marks: 10

Q1. Fill in the blanks: **(3 marks)**

    a. There are _____ months in a year and _____ days in a week.

    b. There are _____ days in February in an ordinary year.

Q2. Answer the following questions: **(3 marks)**

    a. Which months have thirty days?

    Answer:_____

    b. Which month comes between August and October?

    _____

Q3. Arrange the following festivals of India according to the order of their months.   **(3 marks)**

    Holi, Republic Day, Christmas, Guru Nanak Jayanti (usually in November), Eid ul Fitr (usually in August or September), Teej (usually in July).

    Answer _____

    _____

Q4. If the third of a particular month is on a Saturday, which day was the first of the month? _____   **(1 Mark)**

INDIAN MONEY

Game

Learning Objective:

To add money using different currency notes and coins.

Number of Players:

Any

Materials Required:

Playing money, thin cardboard, ruler, pencil, sketch pens, a pair of scissors and paper.

Procedure:

Money Cards

☞ Cut out the playing money from the back of the book.

☞ Make ten money cards as shown above of different amounts.

☞ Put them face down on the table.

☞ Each player has a given set of playing money.

☞ The money cards are turned one by one.

☞ The players have to make that amount by using their playing money.

☞ The first player with the correct arrangement scores two points.

☞ If two players make the correct arrangement at the same time, then both score one point, each.

☞ The game continues till all the money cards are over.

☞ The player with the highest total score is the winner.

Learning Mathematics - The Fun Way

Example:

✯ Suppose the money card turned shows ₹ 652.50.

✯ You may display

Or

Indian Money

- ☞ The first player with the correct arrangement scores two points.
- ☞ The game continues till the last money card is turned.
- ☞ The total score is calculated. The player with the highest total wins.

Skills developed:

Addition of money with and without regrouping.

Donald's Eatery

	Price ₹ प
French fries	
Burger	25.50
Pizza	
Noodles	
Pastry	
Ice Cream	20.25
Soft Drink	

Learning Objective:

To make rate chart and bill using Indian currency.

Materials Required:

Paper, thin cardboard, a pair of scissors, sketch pens and glue.

Procedure:

☞ Using all the stationery items, make your own menu card as shown above and put your own price.

☞ Show the menu card to your own family and your friend's family.

☞ Ask them to select the items.

☞ Make a bill for the two families.

Indian Money

Example:

* ✳ Piklu went to Donald's eatery with his grandparents.
* ✳ All three of them ordered ice cream.

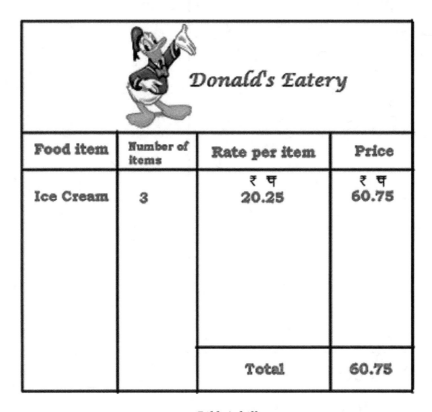

Food item	Number of items	Rate per item ₹ प 20.25	Price ₹ प 60.75
Ice Cream	3	20.25	60.75
		Total	60.75

Piklu's bill

Observation:

Which family had a greater bill by ordering from your menu card?

_____

Result:

The total bill depends upon the price of the food item and the number of items ordered.

Learning Mathematics - The Fun Way

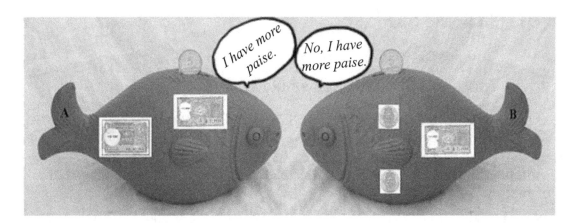

The above two fishes had a fight over the total amount of paise. Fish A has one five-rupee coin, one two-rupee note and one one-rupee note, whereas fish B has one five rupee coin, one two-rupee note and two fifty paisa coins. You be the judge and find out who has more total paise.

Common Error & Correction:

Error	Correction

	₹ प			₹ प
	25.40			25.40
+	14		+	14.00
	25.54			39.40

Tricks & Shortcuts:

When an amount of money does not show any paise, it means there are zero paise. For example, twelve rupees means twelve rupees and zero paise.

WORKSHEET

Q1. Fill in the blanks:

a) One rupee = _____ paise

b) There are ____ fifty paisa coins in a rupee.

Q2. Write the number of notes and coins required to represent the given amount of money.

Amount of money ₹ प									
15.50									
43.75									
59.25									
140.50									

Q3. Look at the price tags and answer the questions given below.

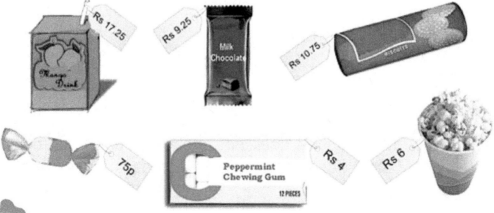

Rs 17.25   Rs 9.25   Rs 10.75   75p   Rs 4   Rs 6

Learning Mathematics - The Fun Way

a) Name two items which can be bought using a ten-rupee note. _____ and _____.

b) Name two items which can be bought using a twenty-rupee note. _____ and _____

c) Which is the minimum currency note required to buy all the items? _____.

Q4. Add the following:

a)

₹ प
17.75
+ 24.00

_____

b)

₹ प
500.00
+ 219.00

_____

c)

₹ प
183.75
+ 111.25

_____

Q5. Subtract the following:

a)

₹ प
76.50
- 39.25

_____

b)

₹ प
782.75
- 418.50

_____

c)

₹ प
543.50
- 226.00

_____

Q6. Mrs.Patel bought a loaf of bread for ₹ 14.50, a jam bottle for ₹ 24 and one dozen eggs for ₹ 48. How much money did she spend in all?

Q7. Rupali's father gave her a fifty-rupee note. She went to a Diwali fair with her brother. They spent ₹ 39.25. How much money should she return to her father?

Q8. Julie works with an artist. She is paid ₹ 38 for completing one painting. How much will she earn for three paintings?

Time: 45 minutes   Maximum marks: 10

Q1. How many twenty-five paisa coins make a rupee? _____
                                                   (1 Mark)

Q2. What is the amount of money shown by the given notes and coins?   **(2 Marks)**

a.

_____

b.

_____

Indian Money

Q3. Look at the price tags of the given items and answer the questions given below. (3 Marks)

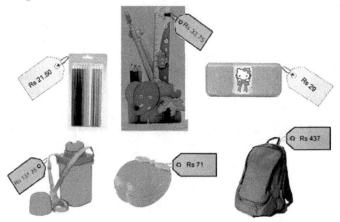

a. What is the total price of one pencil box, two bottles and one bag?

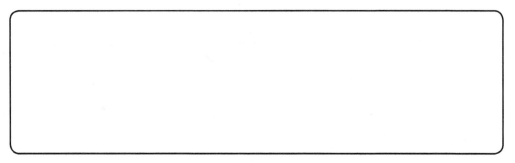

b. Make a cash memo of the items given in part a.

Q4. Andiyappan's family went to Delhi during their winter vacation. They wanted to travel by Metro Rail. His father found out the fares of various stations starting from Central  Secretariat. He wanted to buy four tickets from Central Secretariat to Kalkaji Mandir and gave a hundred rupee note at the ticket counter. How much money will the counter clerk return? **(4 Marks)**

Station	Fare ₹
Lajpat Nagar	15
Kalkaji Mandir	16
Sarita Vihar	18
Tughlakabad	21

 **SMART CHARTS**

Learning Objective:

To draw and interpret a pictograph.

Number of Players:

Any

Materials Required:

Plastic bottle, plastic or rubber ball.

Procedure:

☞ This is an outdoor game. Put a plastic bottle on the ground.

☞ The players stand at a distance of about six metres from the bottle and try to hit the plastic bottle one by one with a ball.

☞ If the bottle falls, then the player draws a smiley face on the table given below.

☞ After ten rounds, the number of smiley faces per player is counted.

☞ The player with the maximum number of faces wins.

Number of children hitting the bottle

Name of the player	Number of times the bottle is hit
Damini	☺ ☺

Skills developed:

Drawing and interpretation of pictograph.

Learning objective:

To record data using tally marks and analyze it.

Materials required:

Paper, pencil, ruler, sketch pens

Procedure:

☞ Make the following table using ruler, pencil and sketch pens.

Colour	Tally Marks	Number
Red		
Blue		
Brown		
Green		
Purple		
Total	_____	

☞ Ask twenty people (friends, relatives and neighbours) their favourite colour.

☞ Put a tally mark (|) against every answer in the correct column.

☞ Use the respective colours for the tally marks.

☞ Add the tally marks and write the number.

☞ Finally add the numbers to verify the total (the total should be twenty).

☞ Note down your observation.

Example:

Colour	Tally Marks	Number
Red	||||	5
Blue			1

Brown		
Green		
Purple		
Total	_____	

Observation:

The most favourite colour is _____.

The least favourite colour is _____.

Result:

Recording data becomes easy with tally marks.

Ice cream Flavour	Number of Children
	?
	4
	3
	2

Amit had an ice cream party for his birthday. He invited 15 of his friends. They had ice creams as per the above chart. How many children had chocolate cone?

Common Error & Correction:

Error   **Correction**

To represent **five**, we use four vertical lines and one sideways line.

WORKSHEET

Q1. Rohit took a survey of favourite fruits in his class. He got the following result:

Favourite Fruit	Number of people	Tally Marks
(apple)	11	
(mango)	15	
(strawberry)	6	
(banana)	8	

a) Put tally marks in the above table.

b) Which is the most favourite fruit? _____

c) Which is the least favourite fruit? _____

d) How many people did Rohit survey? _____

e) What is the difference between the most favourite and the least favourite fruit? _____

Learning Mathematics - The Fun Way

Q2. The following table shows the sale of laptop in one week.

Day	Laptops Sold
Monday	
Tuesday	
Wednesday	
Thursday	
Friday	
Saturday	
Sunday	

a) Find the number of laptops sold on each day of the week.

Monday   _____

Tuesday   _____

Wednessday   _____

Thursday   _____

Friday   _____

Saturday   _____

b) Find the total number of laptops sold during the whole week.

_____

c) On which day were the maximum number of laptops sold?

_____

d) On which day were the minimum number of laptops sold?

_____

e) What is the difference in the number of laptops sold on Friday and Saturday?

_____

Q3. The following table shows the favourite sports of some children.

Sport	Number of Children
⚽	🧍🧍
🏏	🧍🧍🧍🧍
🏀	🧍🧍🧍
🏑	🧍

🧍 = 5 Children

a) How many children play football? _____

b) Which game is played by the maximum number of children?

_____

c) What is the total number of children playing basketball and hockey? _____

Learning Mathematics - The Fun Way

PRACTICE TEST PAPER

Time: 45 minutes   Maximum marks: 10

Q1. The following graph shows the length of different coloured strips. Look at the graph and answer the following questions.   **(5 Marks)**

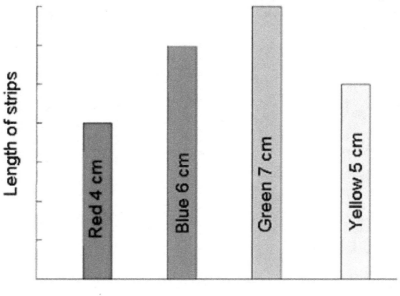

a. Which coloured strip is the longest? _____

b. Which coloured strip is the shortest? _____

c. The length of the green strip is _____ (more/less) than the yellow strip.

d. The length of the red strip is _____ (more/less) than the blue strip.

e. What is the total length of all the strips? _____

Q2. The following table shows the favourite flowers of different children.

Draw the tally marks and answer the questions.   (5 Marks)

Flower	Number of Children	Tally Marks
Periwinkle	8	
Hibiscus	11	
Marigold	8	
Rose	15	

a. Which flower is the most favourite? _____

b. Which two flowers are equally liked by the children?

_____

c. What is the total number of children who like Periwinkle
    and Hibiscus? _____

Learning Mathematics - The Fun Way

MENTAL ARITHMETIC

Q1. Add:

a) 30 + 40 = ____

b) 20 + 60 = ____

c) 60 + 80 = ____

d) 70 + 30 = ____

e) 200 + 100 = ____

f) 300 + 400 = ____

g) 500 + 300 = ____

Q2. Find the sum:

(For adding 9 mentally, first add 10, then subtract 1)

a) 17 + 9 = ___   b) 9 + 25 = ___

c) 48 + 9 = ___   d) 9 + 51 = ___

(For adding 19 mentally, first add 20, then subtract 1)

a) 15 + 19 = ___   b) 8 + 19 = ___

c) 24 + 19 =___   d) 33 + 19 = ___

Q3. Find the difference:

(For subtracting 9 mentally, first subtract 10, then add 1)

a) 23 − 9 = ___   b) 42 − 9 = ___

c) 38 − 9 = ___   d) 85 − 9 = ___

Q4. Find the double of the following numbers:

20___, 50___, 11___, 33___, 25___, 36___, 48___, 15___.

ANSWERS

Number System

Puzzle

213 or 235

Worksheet

A1. a) Number 376; Number name: Three hundred seventy-six

b) Number 527; Number name: Five hundred twenty-seven

c) Number 232; Number name: Two hundred thirty-two

A2. a) 20 b) 400

A3.

A	369	804	212	114
B	100 + 10	873	110	378
C	800 + 9	890	810	801

A4. a) 305, 310, 315, 320, 325, 330, 335.

b) 987, 989, 991, 993, 995, 997, 999.

c) 900, 800, 700, 600, 500, 400, 300.

d) 703, 701, 699, 697, 695, 693, 691.

A5.

School	Number of Students	Sheet of hundred buttons	Sheet of ten buttons	Loose items
A	627	6	2	7
B	418	4	1	8
C	256	2	5	6
D	563	5	6	3

A6. a)

1	3	4	2
4	2	1	3
2	1	3	4
3	4	2	1

b)

3	2	4	1
1	4	2	3
4	3	1	2
2	1	3	4

158

a) Test

A1. a) Nine hundred forty eight

b) Two hundred ninety six

A2. a) 435

b) 762

A3. a) 100   b) 10

A4. a) >   b) 206   c) 380, 400

d) 563 = 5 Hundreds + 6 Tens + 3 Ones

Addition

Puzzle

8	11	15	1
14	2	7	12
3	17	9	6
10	5	4	16

Worksheet

A1. a) 88   b) 37   c) 60   d) 50   e) 75

A2. a) 87   b) 88   c) 85

1	2	3	4	5	6	7	8	9	10
11	12	13	14	15	16	17	18	19	20
21	22	23	24	25	26	27	28	29	30
31	32	33	34	35	36	37	38	39	40
41	42	43	44	45	46	47	48	49	50
51	52	53	54	55	56	57	58	59	60
61	62	63	64	65	66	67	68	69	70
71	72	73	74	75	76	77	78	79	80
81	82	83	84	85	86	87	88	89	90
91	92	93	94	95	96	97	98	99	100

A3. There can be different answers.

45 = 40 + 5, 45 = 30 + 15, 45 = 42 + 3, 45 = 39 + 6, 45 = 35 + 10

327 = 300 + 27, 327 = 301 + 26, 327 = 100 + 227, 327 = 200 + 127

327 = 320 + 7

A4. a) 95   b) 374

A5. a) 900   b) 1000   c) 1000

A6. a) 70   b) 565   c) 444   d) 150

A7. a) 757   b) 863   c) 804   d) 726

A8. 701 people

A9. 950 trees

A10. Answer will vary.

Test

A1. 79

A2. a) 13 + 50 = 63   b) 34 + 16 = 50

1	2	3	4	5	6	7	8	9	10
11	12	13	14	15	16	17	18	19	20
21	22	23	24	25	26	27	28	29	30
31	32	33	34	35	36	37	38	39	40
41	42	43	44	45	46	47	48	49	50
51	52	53	54	55	56	57	58	59	60
61	62	63	64	65	66	67	68	69	70
71	72	73	74	75	76	77	78	79	80
81	82	83	84	85	86	87	88	89	90
91	92	93	94	95	96	97	98	99	100

A3. a) 400   b) 0

A4. 783 books

Learning Mathematics - The Fun Way

A5.

49	+	76	=	125
+		+		+
223	+	359	=	582
=		=		=
272	+	435	=	707

Subtraction

Puzzle

One (Only the dead one remained. The others got scared and flew away).

Worksheet

A1. a) 58 − 23 = 35

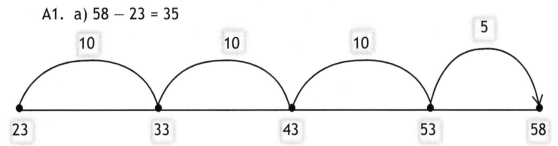

b) 80 − 55 = 25

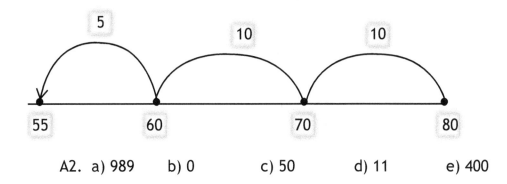

A2. a) 989   b) 0   c) 50   d) 11   e) 400

A3.

1	2	3	4	5	6	7	8	9	10
11	12	13	14	15	16	17	18	19	20
21	22	23	24	25	26	27	28	29	30
31	32	33	34	35	36	37	38	39	40
41	42	43	44	45	46	47	48	49	50
51	52	53	54	55	56	57	58	59	60
61	62	63	64	65	66	67	68	69	70
71	72	73	74	75	76	77	78	79	80
81	82	83	84	85	86	87	88	89	90
91	92	93	94	95	96	97	98	99	100

a) 73 – 20 = 53   b) 88 – 63 = 25

A4.

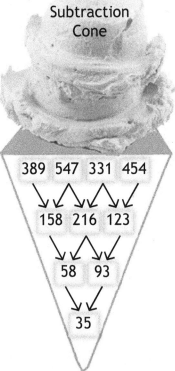

Subtraction
Cone

389 547 331 454

158 216 123

58   93

35

Learning Mathematics - The Fun Way

A5. 109 rakhis.

A6. Pakhi sanctuary, Nine more

Test

A1.

1	2	3	4	5	6	7	8	9	10
11	12	13	14	15	16	17	18	19	20
21	22	23	24	25	26	27	28	29	30
31	32	33	34	35	36	37	38	39	40
41	42	43	44	45	46	47	48	49	50
51	52	53	54	55	56	57	58	59	60
61	62	63	64	65	66	67	68	69	70
71	72	73	74	75	76	77	78	79	80
81	82	83	84	85	86	87	88	89	90
91	92	93	94	95	96	97	98	99	100

a) 88 – 40 = 48   b) 63 – 22 = 41

A2. a) 334   b) 121

A3. a) 411   b) 0

A4. 162 mangoes

A5. Two days

Multiplication

Puzzle

a) 6 × 6 = 36   b) 4 × 9 = 36

Worksheet

A1. a) Addition Sentence: 4 + 4 + 4 = 12

Multiplication Sentence: 3 × 4 = 12

b) Addition Sentence: 2 + 2 + 2 + 2 = 8

Multiplication Sentence: 4 × 2 = 8

Answers

A2. a) 3 × 2 = 6, 3 × 5 = 15, 3 × 9 = 27, 3 × 7 = 21

b) 4 × 6 = 24, 4 × 8 = 32, 4 × 1 = 4, 4 × 2 = 8

c) 6 × 3 = 18, 6 × 6 = 36, 6 × 10 = 60, 6 × 4 = 24

d) 7 × 9 = 63, 7 × 6 = 42, 7 × 3 = 21, 7 × 7 = 49

e) 8 × 5 = 40, 8 × 8 = 64, 8 × 4 = 32, 8 × 2 = 16

f) 9 × 7 = 63, 9 × 6 = 54, 9 × 9 = 81, 9 × 3 = 27

A3. a)

1	2	3	4	5	6	7	8	9	10
11	12	13	14	15	16	17	18	19	20
21	22	23	24	25	26	27	28	29	30
31	32	33	34	35	36	37	38	39	40
41	42	43	44	45	46	47	48	49	50

b

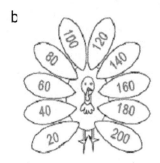

A4. Multiplication Crossword:

¹7	7		²3	3	⁵6
2		⁴2	0		4
	³5	0	0		
⁶3	1	5		⁵9	9

A5. a)

Adding: 5 + 5 + 5 + 5 + 5 = 25

Skip counting: 5, 10, 15, 20, 25

Multiplying: 5 × 5 = 25

b)

Adding: 8 + 8 + 8 + 8 + 8 + 8 + 8 + 8 + 8 = 72

Skip counting: 8, 16, 24, 32, 40, 48, 56, 64, 72.

Multiplying: 9 × 8 = 72

Test

A1. a) 120, 150, 180   b) 75, 50, 25

A2. a) 0   b) 400

A3. a) 312   b) 783

A4.

Adding: 8 + 8 + 8 + 8 = 32

Skip counting: 8, 16, 24, 32

Multiplying: 4 × 8 = 32

Division

Puzzle

Worksheet

A1. a) Number of lollipops = 6

      Number of children = 2

      Number of lollipops each child gets = 6 ÷ 2 = 3

   b) Number of mangoes = 12

      Number of plates = 3

      Number of mangoes on each plate = 12 ÷ 3 = 4

A2.

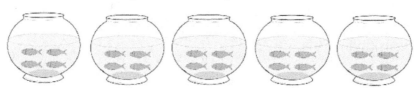

Division fact = 20 ÷ 5 = 4

Division fact = 24 ÷ 6 = 4

A3. a) 81 ÷ 9 = 9, Divisor = 9, Dividend = 81, Quotient = 9

   b) 100 ÷ 10 = 10, Divisor = 10, Dividend = 100, Quotient = 10

   c) 49 ÷ 7 = 7, Divisor = 7, Dividend = 49, Quotient = 7

   d) 21 ÷ 3 = 7, Divisor = 3, Dividend = 21, Quotient = 7

A4. a) 45 ÷ 9 = 5, 45 ÷ 5 = 9, 5 × 9 = 45, 9 × 5 = 45

Learning Mathematics - The Fun Way

b) 28 ÷ 7 = 4, 28 ÷ 4 = 7, 7 × 4 = 28, 4 × 7 = 28

c) 56 ÷ 7 = 8, 56 ÷ 8 = 7, 8 × 7 = 56, 7 × 8 = 56

d) 27 ÷ 9 = 3, 27 ÷ 3 = 9, 3 × 9 = 27, 9 × 3 = 27

Test

A1. a) 9   b) 18 ÷ 2 = 9   c) 6, 18 ÷ 3 = 6

A2. a) 5 × 10 = 50, 10 × 5 = 50   b) 5 × 8 = 40, 8 × 5 = 40

A3. a) 9   b) 2   c) 90

Mirror Halves

Puzzle

Mahatma

Worksheet

A1. a) No   b) Yes   c) No   d) Yes   e) Yes   f) Yes

A2.

   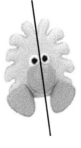

A3.

Test

A1. a) Yes   b) No

A2.

A3. TWEX

A4.

Shapes and Designs

Puzzle

Circle

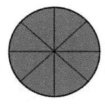

Worksheet

A1. a) S   b) C   c) S   d) C   e) S   f) C & S   g) C & S   h)
     C

A2. a)

Shape	Shape name	Number of sides	Number of corners
	Square	4	4
	Triangle	3	3

	Circle	0	0
	Hexagon	6	6

b)

Shape/Shape name	Number of faces	Number of edges	Number of corners
Cuboid	6	12	8
Cone	2	1	1
Cylinder	3	2	0

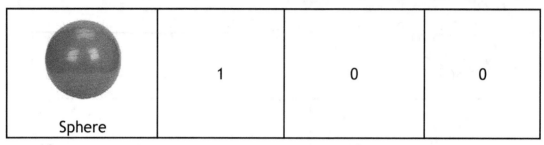			
Sphere	1	0	0

A3.

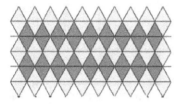 

A4.

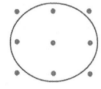 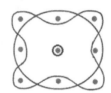

Step 1   Step 2   Step 3

A5. a) Side view of a laptop   b) Top view of a pot

      c) Top view of a plant   d) Side view of a bed

A6.

Test

A1.

Shape	Number of sides	Number of corners
Rectangle	4	4
Pentagon	5	5
Oval	0	0
Cube	12	8

A2. a)

b) Objects made of straight edges will always have corners.

A3.

A4. a)

b)

Patterns

Puzzle

Eye

Worksheet

A1.

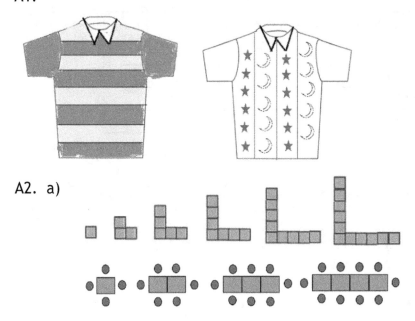

A2. a)

A3. a) 9, 18, 27, 36, 45, 54.

b) 11, 22, 33, 44, 55, 66.

c) 20, 18, 16, 14, 12, 10.

d) 25A, 50B, 75C, 100D, 125E, 150F.

A4. a) 11, 13, 15, 17, 19, 21.

1	13	35	69	47	39	71	51	3
49	17	2	14	50	26	10	15	31
79	43	91	61	83	21	38	87	59
41	25	37	15	95	3	42	45	65
97	9	8	18	46	34	16	11	75
33	63	20	19	53	41	73	85	93
13	55	32	89	99	81	23	57	27
83	69	24	28	40	4	30	53	81
5	71	49	37	67	99	23	79	7

A computer uses only two digits - 0 and 1 to store information.

172

A5. 987JAYANTI: 9875292684
981SAURABH: 9817287224
983JAIDEEP: 9835243337
805SHIKSHA: 8057445742

Test

A1. a) 12, 14, 16, 18, 20. Even
b) 63, 56, 49, 42, 35.

A2. a) A, AB, ABC, ABCD, ABCDE.
b) Apple 1, Banana 2, Cherry 3, Date 4. (Fruits in alphabetical order)

A3. a)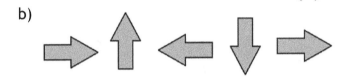

b)

A4. a) Banana (In a dictionary, the names of fruits are arranged in an alphabetical order).
b) Fernandes (In a telephone directory, the names of people are arranged in an alphabetical order).
c) P = 16

Length

Puzzle

Mahesh is taller than Suresh. (Mahesh's height = 103 cm, Suresh's height = 102 cm)

Worksheet

A1. Five buttons

Answers

A2. a) Kilometre, metre and centimetre.　　b) 15　　c) 100

A3. a) Less than one metre　　b) More than one metre

　　c) More than one metre　　d) Less than one metre

A4. Answer will vary. The length of a saree is about five metres.

A5. a) 10 cm　　b) 13 cm

　　c) 7 cm　　d) 2 cm (12 cm − 10 cm)

A6. a) Eden Gardens　　b) Science City

　　c) 16 Km (7Km + 9Km)　　d) Ganga

Test

A1. a) Km　　b) cm　　c) m

A2. Tailor

A3.　Answer will vary

A4. A = 84m, B = 74m, C = 75m Hence, B is the shortest.

A6. Bhopal

Weight

Puzzle

Both are equal (one kg).

Worksheet

A1. a)

　　　3　　　　2　　　　4　　　　1

b)

　　　2　　　　3　　　　1　　　　4

174

Learning Mathematics - The Fun Way

A2. Two oranges

A3. a) Pineapple   b) Apricot   c) 1 Kilogram   d) 500 grams

A4.

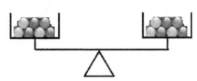

A5. Answers will vary.

A6. a) Tiffin box, calculator b) School bag with books, a pair of adult jeans

A7. a) 1000   b) 14   c) 1

Test

A1. a) gram   b) kilogram   c) gram   d) kilogram

A2. a) more than one kilogram   b) less than one kilogram

c) less than one kilogram   d) more than one kilogram

A3. One kilogram of puffed rice

A4. Salma

A5. a) Railway station   b) Grocery shop

Capacity

Puzzle

Fill the 3-litre bowl and pour the water into the 4-litre bowl. 4 −3 =1. So there will be a gap of 1 litre. Now fill the 3-litre bowl again and pour over the water contained in the 4-litre bowl. It can hold 1 extra litre. Hence, the remaining water in the 3-litre bowl will be 2-litres.

Worksheet

A1. a)

1   2   4   3

2	1	3	4

A2. a) C   b) D   c) 6ml (10 – 4)   d) 100 ml

A3. Robert's fish will be more comfortable since he poured two glasses of water (250 ml + 250 ml = 500 ml).

A4. Ten bottles.

A5. a) 6 to 8   b) half, one

Test

A1. a) litre   b) millilitre   c) millilitre   d) litre

A2. a) less than one litre   b) more than one litre

c) more than one litre   d) less than one litre

A3. a) 5ml   b) 250 ml

A4. Eight bottles

Time

Puzzle

10 O'Clock

Worksheet

A1. a)

Takes seconds	Takes minutes	Takes hours
Blinking of eyes	Cooking kidney beans	Watching a movie
Taking off shirt	Change of traffic lights	Sleeping at night

Learning Mathematics - The Fun Way

b)

Takes days	Takes months	Takes years
Growing plants from the seeds	Change of season	Growing of a child into an adult
	Making of a hand embroidered saree	Construction of a four-storey house

A2. a) 11 O'Clock   b) 1:30   c) 2 O'Clock   d) 6:30   e) 6 O'Clock

A3. a)

10 O'Clock

b)

2:30

c)

8:30

d)

7 O'Clock

A4.

I wake up in the morning at 6 O'Clock.

My school starts at 7 : 30 am.

My school gets over at 1 : 30 pm.

Answers

177

I do my homework at 4 O' Clock.

I go to bed at 9 O'Clock.

Test

A1. a) seconds   b) hours   c) days

A2. a) 12 O'Clock   b) 3:30   c) 9 O'Clock   d) 11:30

A3. a) 12 O'Clock   b) Answer will depend on the year of solving.

      c) 1/1/2021

A4. a)    b)

       8 O' Clock   4 : 30

Calendar

Puzzle

29th February

Worksheet

A1. Clockwise from left:

     April, January, February, July, September, December, November, October, March, August, June, May

A2. a) 15th August   b) 2nd October

      c) 14th November   d) 5th September

Learning Mathematics - The Fun Way

A3. a)

July

SUNDAY	MONDAY	TUESDAY	WEDNESDAY	THURSDAY	FRIDAY	SATURDAY
1	2	3	4	5	6	7
8	9	10	11	12	13	14
15	16	17	18	19	20	21
22	23	24	25	26	27	28
29	30	31				

    a) 18   b) Five

    c) No (In one week, there are seven days, hence, the second Saturday cannot be beyond 14th).

    d) Sunday, Monday and Tuesday.

A4. a, b and c Answers will vary.

    d) January, March, May, July, August, October, December.

Test

A1. a) 12, 7   b) 28

A2. a) April, June, September, November

    a) September

A3. Republic Day, Holi, Teej, Eid ul Fitr, Guru Nanak Jayanti, Christmas.

A4. Thursday

Indian Money

Puzzle

Both have equal money — eight rupees.

Worksheet

A1. a) 100   b) two

A2. This question can have different sets of answers. One set is given here.

Answers

Amount of money ₹ प									
15.50		1			1	1			
43.75	1	1	1	1			2		
59.25	1			2	1			1	
140.50		1				4			1

A3. a) Milk chocolate and toffee or chewing gum and popcorn
 b) Milk chocolate and biscuit   c) Fifty-rupee note

A4. a) 41.75   b) 719.00   c) 295.00

A5. a) 37.25   b) 364.25   c) 317.50

A6. ₹ 86.50   A7. ₹10.75   A8. ₹114

Test

A1. Four

A2. a) 76.25   b) 154.75

A3. a) ₹ 728.50

 b)

		Cash Memo Stationery Shop	
Item	Number of items	Rate per item ₹	Price ₹
Pencil box	1	29.00	29.00
Bottle	2	131.25	262.50
Bag	1	437.00	437.00
		Total	728.50

A4. ₹ 36 (100 − 64)

Smart Charts

Puzzle

4 + 3 + 2 = 9, 15 - 9 = 6 children had chocolate cone.

Learning Mathematics - The Fun Way

Worksheet

A1. a)

Favourite Fruit	Number of people	Tally Marks
	11	ЖЖ ЖЖ I
	15	ЖЖ ЖЖ ЖЖ
	6	ЖЖ I
	8	ЖЖ III

    b) Mango   c) Strawberry

    d) 40 (11 + 15 + 6 + 8)   e) 9 (15 − 6)

A2. a)

Day of the week	Number of laptops sold
Monday	2
Tuesday	4
Wednesday	1
Thursday	2
Friday	3
Saturday	7
Sunday	4

    b) 23 (2 + 4 + 1 + 2 + 3 + 7 + 4)   c) Saturday

    d) Wednesday   e) 4 (7 − 3)

A3. a) 10   b) Cricket   c) 20 (15 + 5)

Test

A1. a) Green   b) Red   c) more

    d) less   e) 22cm

A2.

Flower	Number of Children	Tally Marks
![Periwinkle] Periwinkle	8	☐☐☐ ☐☐☐
Hibiscus	11	☐☐☐ ☐☐☐ ☐
Marigold	8	☐☐☐ ☐☐☐
Rose	15	☐☐☐ ☐☐☐ ☐☐☐

a) Rose   b) Periwinkle and Marigold   c) 19 (8 + 11)

Mental Arithmetic

A1. a) 70   b) 80   c) 140   d) 100
     e) 300   f) 700   g) 800

A2. a) 26   b) 34   c) 57   d) 60
     a) 34   b) 27   c) 43   d) 52

A3. a) 14   b) 33   c) 29   d) 76

A4. 40, 100, 22, 66, 50, 72, 96, 30.

Learning Mathematics - The Fun Way

AFTERWORD

You made it to the end – superb! The journey must have reinforced your knowledge and accelerated your learning. You must have had a fantastic voyage!

This is the end of the book but the search for new ideas goes on. Our efforts to chase away the darkness of ignorance with mathematical lights will not end – neither will the quest of our knowledge.

In **Tagore's** words –

"Who will say the end-word when there is no end – when the clouds end their swirling comes the blissful rain – whatever seems to end does so to the human eye, not to the seamless flow in the universe".

I would like you to share the information and skills you have learned in this book with your brothers, sisters, friends and relatives. Each thing learned will reveal another set yet to be learned. My goal was to lighten the wand of logical curiosity that is already there in your mind and nothing will give me more pleasure than to see the flame burning brighter, stronger and higher. Even if I have succeeded in igniting a small spark I am sure that it will blossom into the mystic aura that depicts purity of knowledge.

CUT-OUT SHEETS

Number Game Cards

Smallest 3-digit number	Smallest 3-digit number using digits, 8, 4 and 6.
Largest 3-digit number using digits, 3, 0 and 1.	Which is greater. 589 or 598?
Place Value of 6 in 369	Largest 3-digit number

✂ Hundreds Block

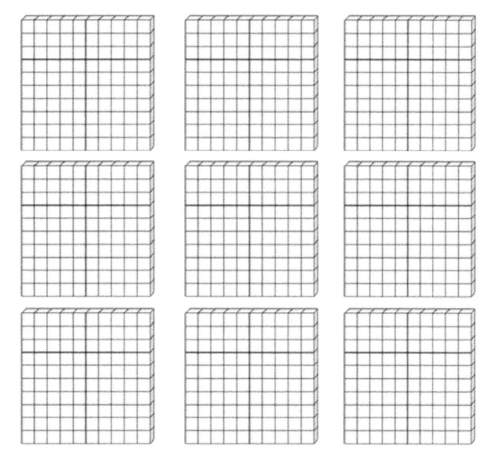

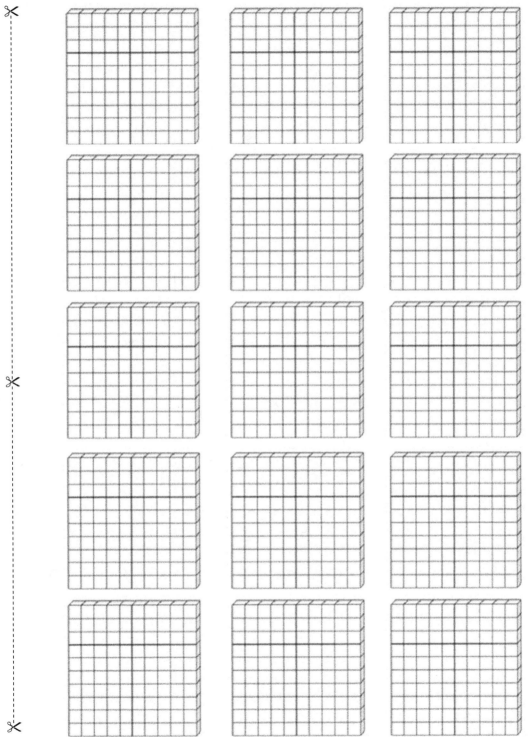

Learning Mathematics - The Fun Way (Cut Out Sheets)

Ones Blocks

Learning Mathematics – The Fun Way (Cut Out Sheets)

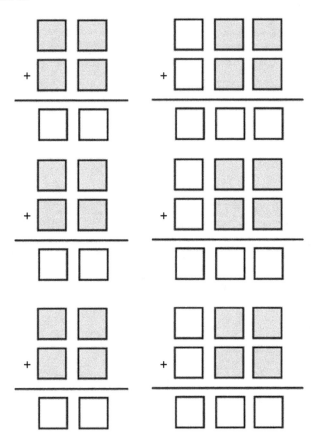

Shape Cards — 1

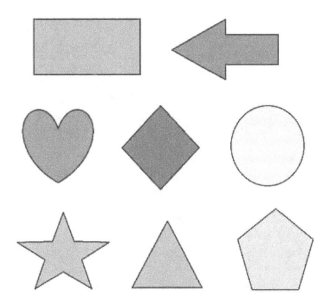

<cut>✂ Shape Cards — 2

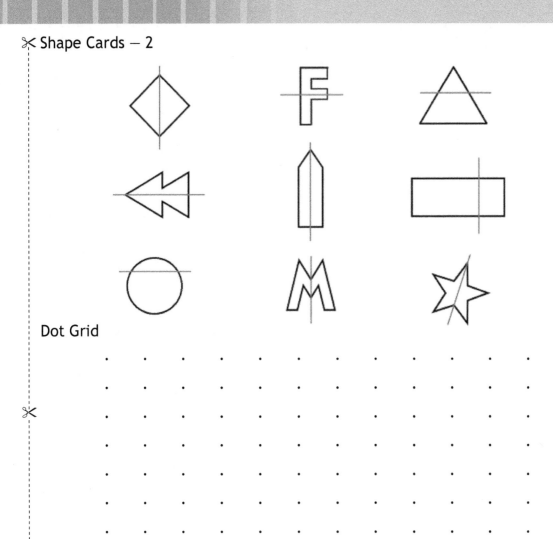

Dot Grid

. . . .
. . . .
. . . .
. . . .
. . . .
. . . .
. . . .

Tangram Set

✂ Tangram Designs

 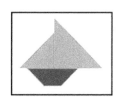   

Envelopes

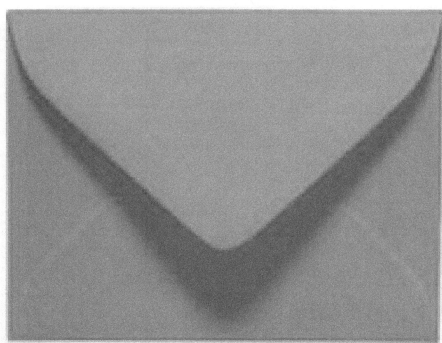

Learning Mathematics - The Fun Way (Cut Out Sheets)

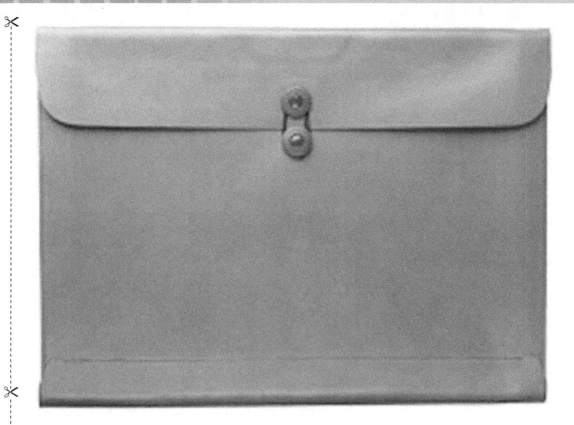

Stamps

Object Cards

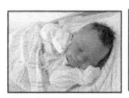  

Learning Mathematics - The Fun Way (Cut Out Sheets)

✂ Weight Cards

More than 1000 kg

About 100 kg	About 10 kg
Less than half kg	About 3 kg
Between half and one kg	Lighter than football

Capacity Cards

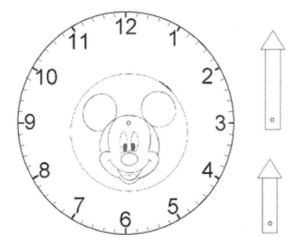

About 4	About 10	About 1
About 11	About 2	About 100
About 30	About 6	About 7

Clock

Learning Mathematics - The Fun Way (Cut Out Sheets)